SOME VITAL QUESTIONS
ABOUT HGH AND AGING

What is Human Growth Hormone?
HGH is a natural chemical produced by the body and is crucial to the growth and maintenance of human tissue.

Why do we need supplements of HGH?
Because HGH, which humans begin to produce at birth, peaks during adolescence, declines in adulthood, and continues to diminish with age.

What is the effect of low levels of HGH?
Scientists are now linking the decrease in HGH to the health-threatening complications of aging.

FIND OUT ALL YOU NEED TO KNOW
ABOUT THIS EXTRAORDINARY AND
CONTROVERSIAL NEW MEDICAL AND
SCIENTIFIC DISCOVERY

HGH: The Promise of
Eternal Youth

Other Avon Books by
Suzanne LeVert

EMERGENCY CHILDCARE
(with Peter T. Greenspan, M.D.)

MELATONIN: THE ANTI-AGING HORMONE

HGH

THE
PROMISE
OF ETERNAL
YOUTH

SUZANNE LEVERT

INTRODUCTION BY GLENN S. ROTHFELD, M.D.

AVON BOOKS ◆ NEW YORK

The ideas, procedures, and suggestions in this book are intended to supplement, not replace, the medical advice of a trained medical professional. All matters regarding your health require medical supervision. Consult your physician before adopting the suggestions in this book, as well as any condition that may require diagnosis or medical attention. The author and publisher disclaim any liability arising directly or indirectly from the use of this book.

AVON BOOKS
A division of
The Hearst Corporation
1350 Avenue of the Americas
New York, New York 10019

Copyright © 1997 by Suzanne LeVert
Published by arrangement with the author
Visit our website at http://AvonBooks.com
Library of Congress Catalog Card Number: 97-93167
ISBN: 0-380-78885-3

First Avon Books Printing: October 1997

AVON TRADEMARK REG. U.S. PAT. OFF. AND IN OTHER COUNTRIES, MARCA REGISTRADA, HECHO EN U.S.A.

Printed in the U.S.A.

WCD 10 9 8 7 6 5 4 3 2 1

Contents

Introduction

Long before Ponce de Leon hunted the Florida coastline searching for the Fountain of Youth, humans have been fascinated with the idea of being young forever. In parts of our present culture, this idea has reached almost manic proportions. We are told by advertisers that being young is the same as being alive. Clothing, music, and entertainment are geared to youthful markets. And modern medicine has adopted a stance clearly favoring the postponement of death at all costs. That means that older people may well have many more years to live than they did a generation ago, but—because chronic illnesses such as heart disease, cancer, diabetes, and arthritis still exist—they fear they will live those years with disease or disability.

It is no wonder that both scientific and popular literature offer countless strategies for staying youthful, or at least for living in a vibrant state of health. Many of these strategies involve lifestyle changes, such as exercise, proper diet, and regular periods of relaxation to lessen chronic stress. The proliferation of natural foods, of yoga and meditation, and of health clubs and workout equipment demonstrates the public move toward life choices which will hopefully influence both well-being in the present, and healthy longevity in the future. Millions of Americans now consume vitamins, minerals, and other nutrients hoping to ward off illness and aging. And where "magic" elixirs were once sold from the back of wagons, we now can go to the doctor for help in growing our hair back, reinvigorating our sex drive, and increasing our childbearing years!

As Suzanne LeVert describes in her book, *Human Growth Hormone: The Promise of Eternal Youth*, hormone therapies are playing a large role in this push toward changing the lifespan. These mysterious substances, many of which are simple bits of protein, have resounding and far-reaching effects on our growth, our moods, our healing, and our susceptibility to illness. The hormones which have become the primary focus of anti-aging researchers are melatonin, dihydroepiandosterone (DHEA), pregnonolone, testosterone, estrogen, and human growth hormone (HGH).

The use of estrogen, of course, has "crossed over" into mainstream medicine, receiving heavy use currently by postmenopausal women. Testosterone is also being prescribed more widely, both in men and, increasingly, in women. Melatonin, DHEA, and pregnonolone are sold over the counter in the U.S., and with the plethora of research becoming available on their effects, we may well be seeing their continued widespread use as both anti-aging substances and as treatments for a wide variety of medical conditions.

Human growth hormone is in a different category. Much of the research on HGH is exciting, provocative, and hopeful. The hormone appears to prompt faster healing, provide energy, and enhance mental and physical function. It has great promise in the possible treatment of degenerative neurologic diseases, and in the wasting of AIDS. It has been shown to boost the immune system and metabolism, and to increase the muscle-to-fat ratio. On the other hand, HGH is difficult to administer, impossible to measure accurately, and extraordinarily expensive. Because much of the current supply of HGH for anti-aging purposes is manufactured out of the country, purity of the hormone is also a concern. Perhaps most importantly, we still don't know exactly how HGH works in concert with other hormones in the body. Because of these issues, HGH remains difficult to obtain, by both interested patients and their doctors.

In a clear, balanced style, Suzanne LeVert describes both the promises and the pitfalls of HGH use. Along the way, she explores much of the physiology related to health and to aging, including the way diet, stress, and exercise affect both hormone levels and general health. She reviews many of the studies which have brought HGH to its current starring role in the longevity quest.

Our bodies contain hundreds of chemical messengers, organized in a remarkably complex web of information and action. Among the most important messengers are the endocrine hormones that travel in our bloodstream, coordinated in their movements and their messages about growth, function, and lifespan. To understand this web, and to influence it, is one of the great medical challenges of our times, and could lead to a true Fountain of Youth. As this book chronicles, human growth hormone plays a central role in the complex process of life and living that we have yet to truly comprehend.

Glenn S. Rothfeld, M.D.
Founder and Director of
Spectrum Medical Arts
Cambridge, Massachusetts
and Clinical Director at
Tufts University School of Medicine

ONE

Human Growth Hormone: The Body's Builder

Q: What is human growth hormone?
As its name implies, human growth hormone, or HGH, is a natural chemical crucial to the growth and maintenance of human tissue. All vertebrates (mammals, birds, reptiles, amphibians, and fish) produce their own type of growth hormone that helps the species grow and develop in specific ways.

Humans begin to produce growth hormone at birth, and varying levels of the hormone circulate in the bloodstream until death. Working with other substances, HGH facilitates the building of bone and muscle as well as the proliferation of cells that allow organs and tissues to grow and repair themselves. HGH levels reach their peak during adolescence, but then drop with each passing year. Many scientists link this decrease to the health-threatening complications of the aging process, including loss of muscle tone and bone strength, increase in body fat, and decreased immune function, among other common problems.

Thanks to recent advances in medicine and technology, scientists can now create a synthetic form of HGH, a substance that the body itself produces only in small quantities. In the fifteen years or so since the introduction of synthetic HGH, researchers around the world have been exploring the effects of HGH on the body in greater depth. As you read this book, you'll see the results of their research provide some important clues to the immensely complex process of growth, development, metabolism, and aging throughout the life cycle.

Q: What exactly does HGH do in the body?
HGH is a protein hormone that performs two primary actions: It promotes *growth* of hard (bone) and soft (muscle) tissues in the body and it influences *metabolism*, or the way the body uses energy. Let's take them one by one.

- **Growth and development**. The most widely studied property of HGH is its effect on growth and development during childhood and adolescence. HGH facilitates the ability of muscle, bone, and other cells to use amino acids—the essential building blocks of protein—to grow and proliferate. Because it has an effect on protein metabolism and water retention, it may also help keep skin taut and supple. Without growth hormone, children cannot reach their full adult height and body type potential. In adults, HGH deficiency results in the loss of lean body mass (organ and muscle tissue), a decrease in muscle strength and bone density, and an increase in body fat.

- **Metabolism.** Metabolism is a word scientists use to describe the sum of all the chemical processes in the body, primarily those that involve the use of nutrients to help the body grow, use energy, and release waste. HGH plays a crucial role in metabolism. Its presence first stimulates the release of fat from body tissues and then triggers heart and muscle tissue to use fat (instead of blood sugar) as energy. Without HGH, the body stores fat much more readily. This aspect of HGH offers widespread—indeed potentially lifesaving—benefits to the body. Current research links excess fat stores to many chronic and debilitating diseases of old age—including heart disease, diabetes, and certain types of cancer. By restoring youthful levels of HGH, scientists hope to diminish the risk of developing the conditions most aging Americans face.

Q: Does human growth hormone play any other roles in the body?

Research indicates that HGH may help repair and maintain many different types of body tissue. Recent avenues of research include the use of HGH to treat severe burns, to help repair injured heart muscle involved in a disease called cardiomyopathy, and to facilitate the repair of damaged nerve cells in amyotrophic lateral sclerosis (ALS), a degenerative disease of the nervous system.

As we examine the role of HGH, however, it's important to consider the complexity of the endocrine system—the system of the body responsible for the production and activity of HGH and other hormones. Each of these chemicals works in coordination with one another, as well as in coordination with the nervous system, another highly intricate network. The two systems work in such close coordination with one another, in fact, that the term "neuroendocrine system" is used to describe their activities. Scientists must complete much more research before they fully understand HGH's impact on the body or how to manipulate the hormone in order to increase human health and longevity.

Q: Where does HGH come from?

Just this one answer will show you the complexity and interconnection of the neuroendocrine system. First, the levels of hormones in the body are never steady or static, but instead constantly change in response to signals from a variety of sources within the body. It might help for you to think of it this way: The pituitary and other endocrine glands react to hormonal changes in the blood in much the same way that a thermostat reacts to temperature changes. The glands do not constantly secrete hormones but instead rely on the presence or absence of hormones in the blood to turn their secretions on and off. If there is not enough of one type of hormone circulating in the blood, the endocrine gland makes more. If too much hormone is present, the glands stop producing it, leading to lower blood levels. The liver also plays an important role in this system, since it removes hormones from the blood and destroys them when levels are too high.

HGH is one of many hormones secreted by the pituitary gland, an organ located at the base of the brain. The pituitary gland releases HGH after receiving a message from the hypothalamus, a part of the brain responsible for a host of neuroendocrine actions. The hypothalamus sends this message by way of another hormone, aptly named growth-hormone-releasing factor or GRF. Once the body has enough HGH in its system, the hypothalamus then sends another hormone, called somatostatin or GH-inhibiting factor, to stop the pituitary from releasing more HGH. Similar feedback systems control the activities of most hormones as they run through the entire body, and are highly effective in maintaining appropriate levels of many different hormones in the body.

Q: What other hormones does the pituitary gland secrete?

Often called "the master gland," the pituitary gland secretes several essential hormones, many of which work in close harmony with HGH. Thyroid-stimulating hormone (TSH), for instance, controls the release of hormones from the thyroid. Thyroid hormones both help to synthesize HGH in the pituitary as well as act with HGH to regulate the body's metabolic rate. Before puberty, the thyroid may also stimulate the thymus gland to produce immune system cells that protect our bodies from disease.

In coordination with the hypothalamus, then, the pituitary helps to direct body temperature, muscle tone, sexual development and reproduction, respiration, water and sodium balance, immune activity, appetite and thirst, fat metabolism, protein synthesis, pain perception, and several aspects of emotion, memory, and learning. HGH is most directly involved in muscle tone, protein synthesis, and fat metabolism, but its presence may help the other hormones perform their functions properly and vice versa.

Q: When does the body produce HGH?

Your body receives small bursts of HGH from the pituitary about every four hours until you fall asleep—and that's when you receive your biggest shot of this powerful hormone. During Phases 3 and 4 of sleep, the deepest and heaviest stages, you obtain fully 70 percent of your entire daily dose of HGH. In fact, scientists believe that the need all humans have for sleep stems not just for the rest it provides, but also for the HGH the pituitary provides to muscles, bones, and other tissues in order to maintain and repair themselves. Melatonin, a hormone that helps prepare your body for sleep, also diminishes with age, and scientists believe that there may be a connection between the loss of sleep, the loss of HGH, and the aging process.

Q: How do doctors measure the amount of HGH in the body?

Once the pituitary gland secretes HGH, the hormone travels to the liver where a chemical process converts it into substances called insulinlike growth factors or IGF. There are several types of IGF but two are directly linked to HGH: IGF-1, which works directly with HGH to promote growth and regulate metabolism, and IGF-2, which appears to act more independently of HGH to maintain the health of nerve tissue. Unlike HGH, which is released in periodic bursts, levels of IGF remain fairly steady throughout the day. For that reason, scientists often use IGF levels to evaluate how much

HGH the pituitary produces, even though the exact function of these factors remains under investigation.

Q: What happens if the body doesn't produce the right amount of HGH?

That depends on when the problem occurs. If a child produces too much HGH, a condition known as gigantism, marked by excess height and weight, could result. If a child doesn't produce enough HGH, on the other hand, he or she could suffer stunted bone growth and stature, a condition known as dwarfism. Scientists first developed synthetic HGH to treat HGH-deficient children, and even today, that remains the hormone's primary medical use.

Some adults also suffer from HGH-related health problems. If the adult body secretes too much HGH, a condition called acromegaly may result: The hormone stimulates bones to grow abnormally, causing the development of deformities in the spine, joints, digits, and face. Excess HGH may also trigger the development of diabetes and cause fluid retention, joint pain, and other symptoms. IGF-1, the growth factor related to HGH, also carries with it some potential risks when present in excess. IGF-1 acts to stimulate cell division, especially in the breast and colon. If any cancer cells are stimulated by IGF-1, the presence of IGF-1 may help cancer to grow and spread. IGF-1 may also play another role in cancer by preventing cells from self-destructing as they normally do (a process called "apoptosis"). Apoptosis often leads to precancerous conditions such as hyperplasia of the cervix and benign prostatic hypertrophy. Many scientists worry that careless use of HGH as an anti-aging therapy may result in the development of these potentially serious side effects.

Some adults produce too little HGH, either because they were born with an HGH deficiency problem or because a tumor affected the pituitary gland in adulthood. Too little HGH in adults causes muscle and bone loss, weakness, and increases in body fat, especially in the abdominal area. These symptoms may trigger the development of osteoporosis and heart disease, among other conditions. For this reason, the FDA approved the use of replacement HGH for adults with clinical HGH deficiency in 1996.

Researchers, struck by the widespread problems with tissue maintenance and metabolism that results from clinical HGH deficiency, are now exploring the hormone's role in the aging process. In fact, some scientists are asking if aging might be merely a hormonal disorder caused by a lack of HGH, a disorder that we could correct by replacing HGH.

Q: Why does our supply of HGH run out as we get older?

At this point, scientists just aren't sure why supplies of this essential hormone decrease so dramatically: At the age of sixty, for instance, most of us produce only about 25 percent of the levels we produced in our twenties. It might be that the pituitary gland itself begins to shrink and is therefore unable to produce enough HGH. Older people also have difficulty falling into a deep sleep, which could prevent the pituitary from releasing its largest burst of HGH. Or the liver could fail to convert enough

HGH into IGF-I, and thus the hormone may be unable to do its job effectively. Scientists continue to look for answers.

As you'll see in Chapter Three, other important hormones crucial to health and vitality also diminish with age, including estrogen, testosterone, melatonin, and DHEA. We'll discuss the impact these hormones might have on the production and action of HGH, as well as their own connection to the aging process.

Q: Where does the supply of HGH come from to treat growth disorders and other conditions?

Until the late 1980s, scientists derived HGH from cadaver pituitary glands. Expensive to secure and able to yield only a limited supply—it takes fifty cadavers to provide enough HGH to treat a single child for a year—cadaver-derived HGH proved too scarce for either treatment or research purposes. In addition, the cadaver-derived HGH also exposed some patients to a rare and fatal brain virus known as Creutzfeldt-Jakob disease. This virus may lay dormant in brain tissue for many years but, once activated, it causes rapid degeneration of brain cells resulting in dementia, muscle wasting, and various uncontrolled movements. Most people with the disease die within a year after symptoms appear.

In order to prevent the inadvertent spread of Creutzfeldt-Jakob disease, as well as increase supplies of HGH to children suffering from growth disorders, scientists developed a synthetic version of the hormone. In October 1985, Genentech, Inc. of San Francisco created the first recombinant human growth hormone (rHGH) to receive approval from the Federal Drug Administration. The company called its product Protropin. About a year later, the pharmaceutical firm of Eli Lilly released a second form of rHGH, called Humatrope. In 1996, another company, Serono Laboratories, created a new brand, called Serostim. Today, other biotechnology companies, here in the United States and around the world, are in the process of creating other brands of HGH.

Q: How is recombinant HGH made?

Scientists make rHGH using molecular biological techniques commonly known as bioengineering. They combine genes of different organisms to form a hybrid molecule. Composed of strands of information-packed DNA (deoxyribonucleic acid), genes "tell" cells how to behave and function. To make rHGH, technicians insert the gene carrying HGH's genetic information into a special plasmid, a circular piece of DNA derived from one organism.

Genentech and Eli Lilly both use a strain of E.coli bacteria as their plasmid. Once they insert the HGH into the E.coli bacteria (after carefully treating it to disable its disease-causing characteristics), the HGH DNA can work as a kind of template. When the bacterium multiplies, it does so carrying HGH DNA—in effect, it *becomes* HGH—and thus every time it multiplies it creates more hormone. (When they make Serostim, the Serono lab technicians use mammalian rather than bacterial cells as their plasmid, but the process is just the same.)

Q: I've also heard of a bovine growth hormone. Is it related to rHGH?

Recombinant bovine growth hormone, or rBGH, is a synthetic version of the hormone that all cattle produce naturally. Like HGH in humans, BGH promotes growth and development of muscle and other tissue. When ranchers implant rBGH and other hormones (specifically the sex hormones estrogen and progestin) into their steers and heifers, the cattle grow significantly faster than nonimplanted animals. Because of that, the animals reach market weight earlier and have more lean meat and less fat. Implants reduce the amount of feed necessary to produce each pound of edible lean meat. And cows implanted with rBGH produce more milk—between 5 and 10 percent more milk—without proportionately increasing production costs. In November 1993, the FDA approved the sale of milk from cows injected with rBGH in order to increase milk production.

Q: Is rBGH safe?

That question remains one of the most controversial in the field of environmental medicine. The food industry and most mainstream medical establishments, including the U.S. Food and Drug Administration and the National Institutes of Health, claim that meat and dairy products from rBGH-treated animals are perfectly safe for human consumption.

Several studies show, however, that these products may not be so safe—for either animals or humans. Veterinarians link the use of rBGH to painful udder infections in cows (mastitis), infections that may contaminate the milk produced with clots of bacteria and the antibiotics used to treat the infection.

Furthermore, rBGH triggers the production of IGF-1 in cows, just as it does in humans. In fact, there are close biological similarities, including an identical chemical structure, between bovine and human IGF-1. rBGH-laden milk may cause an excess of IGF-1 in people who drink it, and thereby may contribute to the development of breast, colon cancer, and other cancers. As you might suspect, the ultimate, long-term health effects of rBGH deeply concern some scientists, who also worry about the safety of using human growth hormone— which may cause the same problems in humans— as an anti-aging supplement.

Q: How does rHGH work in the human body?
The structure of rHGH is identical to that of the natural hormone and so works the same way on body tissues. As mentioned, scientists first developed rHGH specifically to treat growth hormone deficiencies in children. In 1996, the FDA approved its use for HGH deficiencies in adults as well as for use in patients suffering from AIDS wasting, a side effect of infection with the Human Immunodeficiency Virus (HIV).

Although the development of rHGH made the hormone more available to researchers and physicians alike, it remains an expensive and relatively scarce commodity. For that reason, large-scale and long-term studies on the uses and side effects of rHGH are still in their early stages. A current and highly publicized avenue involves rHGH and the aging process. Scientists are considering the hormone's potential as a treatment for various side effects and complications of aging including the following:

- **Osteoporosis**. This bone disease, resulting in weak and porous bones, may benefit from the ability of HGH to build bone density. A 1990 ground-breaking study published in the *New England Journal of Medicine*, for instance, showed a 1.6 percent increase in bone density in the spines of older patients on HGH therapy. As we'll discuss further in Chapter Four, bone metabolism is a highly involved process that requires sufficient nutrients in addition to HGH. Bone health and integrity also benefit from strenuous exercise, which HGH can help older people perform by strengthening their muscles.

- **Loss of muscle mass**. People with HGH deficiency tend to lose lean muscle and replace it with fat. Replacement HGH may counter that effect. The same study in the *New England Journal of Medicine*, for instance, showed that treatment with rHGH increased muscle mass by about 8.8 percent and decreased body fat by 14.4 percent. HGH works to promote protein synthesis, the process by which body tissues like muscles and cell membranes are formed.

- **Heart disease**. Although we tend to forget this fact, the heart is the body's hardest working muscle, and it requires sufficient HGH to maintain its strength throughout the life cycle. Doctors have achieved some success in using HGH to treat patients suffering from cardiomyopathy, a chronic, degenerative disease of the heart muscle. Another avenue of research involves HGH's role in the metabolism of cholesterol, a lipid (fatlike substance) linked to coronary artery disease and atherosclerosis.

- **Obesity and related problems**. HGH's role in fat metabolism may help to decrease weight-related effects on the cardiovascular system and throughout the body. We know, for instance, that the risk of heart disease is highest among men and women who carry excess weight in the abdomen and trunk— just the weight most people who use replacement HGH lose first. Losing fat and gaining muscle will also help reduce the risk of diabetes and certain types of cancer.

- **Breathing difficulties**. Not only do we need our lungs to be clear in order for us to breathe properly, we also need strong chest, stomach, and back muscles. Physicians at the University of North Carolina in Chapel Hill recently improved the breathing capacity of emphysema patients by treating them with growth hormone. HGH treatment strengthened their chest muscles, enabling them to take in more air with less discomfort.

- **Organ failure**. HGH may help increase the lean mass of our internal organs, including the heart, liver, kidneys, and endocrine glands—all of which tend to deteriorate as we age. By increasing their mass, HGH may also help improve their function.

- **Surgery and other trauma**. In addition to helping build muscle and bone tissue, HGH also works to heal skin wounds of all kinds, including those caused by surgery and by severe burns.

- **Depleted immune system**. As mentioned, the release of HGH stimulates the secretion of thyroid hormone which, in turn, acts on the thymus to generate immune system cells. Furthermore, HGH replacement may help to prevent the thymus gland from atrophying due to aging, thus allowing this important organ of the immune system to continue producing healthy immune cells.

- **AIDS wasting**. One of the most devastating side effects of AIDS is a condition known as "wasting." In fighting the disease, the body feeds off itself, sapping its strength and stamina. In fact, many people with AIDS die not of a specific infection, but because their bodies simply waste away. rHGH treatment helps people suffering with this condition to gain weight and muscle mass, and therefore stay stronger and healthier longer. Therapy with rHGH may help people suffering from similar disorders as well: A study released at the seventy-seventh annual meeting of the Endocrine Society in Washington, D.C., for instance, showed that patients suffering from Crohn's disease, a chronic inflammation of the bowel that makes absorbing enough nutrition difficult, benefited from treatment with rHGH. The hormone increased body weight, lean body mass, body cell mass, and total body water.

Q: I've heard that rHGH will improve my sex life. Is that true?

To date, no conclusive studies prove that claim, nor do scientific explanations exist for how or why rHGH would affect sexual performance or satisfaction. Nevertheless, many people who have used HGH claim they experience an increase in libido. Part of the reason for this helpful side effect could be the improvement in general muscle tone and energy levels rHGH engenders. In addition, there is a direct link between the pituitary hormones such as HGH and the sex hormones produced by the ovaries and testes, so that increasing HGH levels also increases estrogen and testosterone levels.

Q: Who should take rHGH?

That's a good question, and one that no one can answer with any certainty at this point. To date, the FDA officially approves the use of rHGH for just three conditions: HGH deficiency in children and adults (discussed in depth in Chapter Two) and AIDS wasting (discussed in Chapter Seven). Because recent studies show that the majority of people lose HGH as they age—and thus also lose its powerful effects on the repair of cells and the metabolism of fat—some physicians are hoping that the FDA will make the hormone more available and affordable to older people with age-related HGH deficiencies.

Currently, however, most doctors only prescribe rHGH to patients with one of the three conditions described above. Some doctors will also use it to treat severe burns, cardiomyopathy, or other serious medical conditions, although such treatment is still considered controversial.

Q: How do I get rHGH?

At this point, rHGH is available only by prescription, and most doctors will not prescribe it as an anti-aging therapy. In "Resources, Reading, and References" on page 204, we list two associations that can put you in touch with reputable physicians who might be willing to prescribe it to you. Before you start taking rHGH, a qualified health professional must first assess your physical health to see if rHGH is appropriate for you, then arrange for you to receive a supply of rHGH. Because the FDA has not approved rHGH for anti-aging purposes, the doctor will probably order your supply of the hormone from an American company that manufactures the drug in another country, usually Mexico. The prescribing doctor will then teach you how to inject yourself with the hormone, either every day or three times a week. Treatment is expensive: A year's supply of rHGH can cost you approximately thirteen to twenty thousand dollars a year.

Be warned: There is currently an extensive black market for the drug, especially among athletes, offering untested rHGH of unknown quality and safety. Reports of black market stock from Eastern Europe that contains old supplies of natural growth hormone tainted with Creutzfeldt-Jakob disease are particularly alarming. You should only purchase rHGH from a reputable physician who will carefully evaluate your progress and monitor any side effects.

Q: Why is rHGH so difficult to obtain?
Hormones are extremely powerful substances that trigger a host of actions within the body. If your body is truly deficient, then replacing just the right amount of a missing hormone makes sense. Diabetics do it quite routinely with the hormone insulin, for instance. We also do it for women who run out of estrogen at menopause, thereby helping them avoid the pitfalls the loss of this essential hormone may cause. (Of course, many physicians have serious reservations about estrogen replacement, too, since estrogen may trigger breast and endometrial cancer, as we'll discuss further in Chapter Eight.)

The controversy over HGH replacement is even more intense. Both the Federal Drug Administration and much of the mainstream medical community feel that, at this point, there is simply too much we don't know about rHGH: How much is too much? How little is too little? What are the long-term risks and side effects? Do the benefits outweigh the risks? And are we sure we know exactly what the benefits truly are for the average healthy older man and woman?

As more scientists conduct more studies over a longer period of time and on more patients, the medical community may discover that the benefits of replacing this hormone will indeed outweigh its risks and side effects. As scientists continue to refine bioengineering techniques, the cost of producing rHGH may decrease as well.

Q: You mention side effects. What are they?
There are indeed some serious side effects connected with rHGH. First, excessive levels of HGH

in the adult body may result in acromegaly, which involves excessive growth of the hands, feet, and face. Another serious risk involves HGH's effects on blood sugar. As we'll discuss further in Chapters Two and Three, the body usually uses glucose—circulating blood sugar derived from recently digested carbohydrates—as energy. HGH, however, releases fat into the bloodstream and then triggers cells to use fat instead of glucose. Although the immediate effect is to aid weight loss, over time, this action may produce abnormally high blood sugar levels, a condition known as hyperglycemia or, eventually, diabetes, in which the body no longer metabolizes glucose properly.

Another fear is that the ability of HGH and IGF-1 to stimulate cell proliferation could result in the growth and spread of cancer cells. We know that people with acromegaly—the production of too much HGH and IGF-1 usually triggered by a pituitary tumor—run higher risks of developing breast and colon cancer. Although no cases of cancer have yet been directly linked to rHGH used for anti-aging purposes, many scientists are voicing concerns that such a possibility exists. In addition to these serious complications, treatment with rHGH often results in fluid retention and swelling, muscle aches and pains, and carpal tunnel syndrome.

Please note that completed studies on rHGH replacement are extremely limited, and anecdotal reports about its effects vary greatly. Some people claim that rHGH treatment allows them to feel younger, stronger, and more energetic than ever before, while others drop out of treatment because of its expense, its side effects, or both. It appears that much depends on determining the perfect dose of rHGH for each individual so that the amount given is just enough to make up for the age-related loss, but not enough to produce side effects.

Q: Is there any way to get the benefits of rHGH without taking the hormone itself?
In Chapter Eight, we describe several supplements, including amino acids, that some researchers believe can help trigger the release of HGH from the pituitary. In Chapter Nine, we talk about lifestyle changes you can make to increase your lifespan and your vitality. We show you how to eat a healthier diet, sleep more deeply, and make exercise a regular part of your daily life.

Q: I've heard about other substances that supposedly have anti-aging effects, such as melatonin and DHEA. Are they related to HGH?
Like HGH, melatonin and DHEA are hormones that have wide-ranging and crucial effects on the body. And also like HGH, natural supplies of melatonin and DHEA dwindle as we get older. Many chronobiologists—scientists who study the body's rhythms through the life cycle—believe that the endocrine system lies at the center of the body's life cycle rhythm and thus holds the key to the aging process. If that theory holds true, then replacing these hormones to their youthful levels may be the best way to turn back the hands of the "aging clock."

Q: What exactly is DHEA?

DHEA (dehydroepaidrosterone) is a hormone produced by the adrenal glands, which are located on top of each kidney. The most abundant steroid hormone in the bloodstream, DHEA appears to have significant anti-obesity, anti-aging, and anti-cancer effects. In fact, DHEA is sometimes called the "mother of hormones" because the body uses it to manufacture many other hormones, including sex hormones (estrogen, progesterone, testosterone) and stress hormones (cortisol, norepinephrine). In addition, certain body cells also have DHEA-specific receptors, which means that DHEA directly affects body tissue and physiology as well.

We lose DHEA, and its benefits, as we age: By the age of fifty, most women's DHEA levels decrease to an average of 50 percent and men's to 43 percent. By age seventy, DHEA levels for women average 31 percent and men 2 percent. As people near death, DHEA blood levels for most people will be between 0 and 5 percent. In recent years, DHEA supplements have been available to replace the loss that occurs with every passing year.

Q: What effects does DHEA have on the body?

Like HGH, DHEA works on several parts of the body and affects many different aspects of physiology. Here are a few examples.

- **Stress and mood.** In a study performed at the University of California Medical School at La Jolla and reported in the June 1994 issue of the *Journal of Clinical Endocrinological Metabolism*, a group of men and women age forty to seventy were given 50 mg. of oral

DHEA for six months. At the end of the study, 84 percent of the women and 67 percent of the men reported a remarkable increase in their physical ability to handle stress. In addition, a group of scientists presented evidence at the New York Academy of Sciences Meeting in June 1995 that DHEA had both anti-depressant and cognition-enhancing properties.

- **Cardiovascular disease**. In another study, reported in a 1986 issue of the *New England Journal of Medicine*, men fifty to seventy-nine years of age given DHEA experienced a 48 percent reduction in death from cardiovascular disease and a 36 percent reduction in death from any cause. Other studies show that an increase of DHEA levels stops the development of atherosclerosis (buildup of arterial plaque) and reduces the risk of death from cardiovascular disease.

- **Alzheimer's disease**. In a study at the Laboratory of Biochemical Genetics in Bethesda, Maryland, a group of researchers found that Alzheimer's patients had 48 percent less DHEA in their bodies than a healthy group of the same age. Replacing DHEA may help diminish the effects, or even prevent or delay the onset of this devastating disease.

- **Immune system**. At Kentucky University's Sanders-Brown Center of Aging, researchers found that the antibody response of mice declined with age, yet DHEA significantly reversed this immune insufficiency. In another study with mice, the production of autoantibodies—cells that attack the body's own tissues as in rheumatoid arthritis—was significantly restrained by treatment with DHEA.

Q: And what about melatonin?
Melatonin is a hormone secreted by the pineal gland, a tiny pine-cone-shaped endocrine gland located at the base of the brain. Nicknamed "the chemical expression of darkness," melatonin (like HGH) is produced almost exclusively at night. And, also like HGH and DHEA, our supply of melatonin dwindles as we age—just when we need it most. The following lists some of the benefits of this powerful hormone.

* **Master hormone**. Research emerging from laboratories around the world indicates that melatonin acts as a prime stimulator of all hormonal activity. For instance, the blood level of melatonin appears to trigger the adrenal glands and gonads to increase or suppress the secretion of female and male sex hormones. Its presence or absence also affects the production of pituitary gland hormones, including human growth hormone.

* **Immune system booster**. In animal trials and human tests, melatonin stimulates the production of antibodies, the body's first line of defense against infection. It also restores production of natural killer cells, a critical part of the immune system and one that appears to decline with age.

- **Super antioxidant**. According to current research, melatonin may be one of our most effective weapons against free radicals, substances that appear to be among the human body's most pervasive and damaging enemies. Free radicals are highly reactive molecules that can alter our DNA (the genetic code regulating cell growth and activity), damage proteins (the main building material for muscles, blood, skin, nails, and organs), and disrupt other body constituents. Scientists have linked free radical damage to myriad age-related diseases, including atherosclerosis and cancer, as well as to the cosmetic effects of aging, such as the wrinkling of the skin and graying of hair. The more free radical activity that takes place, the greater our need for sufficient amounts of HGH to repair the resulting damage to cells and tissues.

Q: Why are we just hearing about HGH and other hormones now? Is it just another fad?

The first question is simple to answer: We needed advanced technology to first identify, then isolate, these hormones for study. Humans produce hormones in very small quantities (one trillionth of a gram and under) and chemical reactions, like protein synthesis, take place in nanoseconds (one billionth of a second or less). Only by using recently developed medical and research tools could scientists begin to accurately assess the potential for HGH and other hormones on our health.

The second question is a little more complex: It is likely that the popularity of HGH, melatonin, and DHEA will rise and fall depending on a host of factors, including the release of study results, the expense and availability of the supplements themselves, and the amount of media attention they receive. There is little doubt, however, that the study of hormone replacement as a "treatment" for aging will remain at the forefront of science. We hope that this book answers some of your questions about this exciting new frontier of medicine.

∽

TWO

Growth and Metabolism Through the Life Cycle

Q: How does my body know how much to grow?

From the moment of conception and throughout your life, much of the responsibility for your growth and development rests largely on two factors: genetics, the information you inherit from your parents, and nutrition, the quality and quantity of nutrients you provide your body. Other factors also influence the way you grow and develop, of course, including the amount of physical activity you perform, your exposure to infectious disease, and the amount of nurturing—both physical and psychological—you receive from your family during your infancy, to name just a few.

Q: Both my husband and I are pretty short. Does that mean our child is destined to be short, too?

Answering that question requires providing a little information about genetics, the science of heredity.

When you and your husband conceive a child, you each donate certain information about how that child will grow and develop. That information comes in the form of chromosomes, threadlike structures in the cell's center. Each chromosome is made up of a double strand of twisted DNA (deoxyribonucleic acid), and along the length of each strand of DNA lie genes, chemicals that contain instructions about physical or physiological traits that they impart to cells.

When your husband's sperm fertilizes your egg, a single new cell—called a zygote—is formed. This zygote contains twenty-three chromosomes from you and twenty-three chromosomes from your husband. The genetic information contained on these chromosomes tells the zygote how to grow and function.

Genes appear in pairs—you contribute one set of genes copied directly from your chromosomes and your husband contributes the other copied from his. Each pair is located at the same position on each of the chromosomes. Sometimes the genes are identical, which means that the child will definitely express that trait. If you and your husband both contribute the gene for blue eyes, for instance, you'll soon be looking into your child's "baby blues." In other instances, you'll contribute one form of the gene, and your husband another. In that case, whichever gene is dominant will carry the day. If you have blue eyes and your husband has green and your baby is born with green eyes, we can assume that your husband carried the dominant gene for eye color. Geneticists call this process "simple inheritance."

Height, on the other hand, is a trait that depends

on many genes as well as environmental influences. Several different genes make contributions to your child's height, as will nutritional and other environmental factors. Geneticists call this process "multifactorial inheritance." What will the genetic information you and your husband pass on to your child mean to your child's ultimate height? That's a question scientists cannot yet answer with any certainty: There are simply too many variables. Although you're both on the short side, for instance, you or your husband may contribute enough genes for tallness that your child might just shoot up past you in terms of height.

It's more likely, though, that your child will be around your same height. Pediatricians have developed this simple formula to help you predict your child's adult height: Combine your and your husband's height together in centimeters. If your child is a boy, add 13 centimeters to that figure and then divide by two. If you have a daughter, subtract 13 centimeters from your combined height and divide by two. A child who grows normally usually falls within 10 centimeters of the resulting measurement.

Q: I'm not sure I understand. If life begins with just a single cell, how does that cell become a whole human being that knows how tall to grow?

A thorough answer to that question would fill several books, but here's a short and simple overview: The zygote—the original cell that formed when your husband's sperm fertilized your egg—divides by a process called meiosis. Every time the cell divides, it carries with it the same twenty-three pairs of chromosomes you and your husband donated. Thus each new cell contains the same basic genetic information.

Development then takes place in three different ways: growth (size increase), morphogenetic movement (the shaping of patterns and forms), and differentiation (the change from general to specific structures). In humans, growth occurs as cells continually divide to form the component cells of all the tissues of the body. Then cells move to the region in which they belong (liver cells to the liver, etc.). During and after this movement, cells become different from one another in chemical composition and structure. This cell differentiation may occur in groups of cells to form tissues (such as muscles and nerves), which in turn form organs (such as the heart or brain). And the differentiation process continues: Some endocrine gland cells become pituitary gland cells, for instance, while others become thyroid cells, and some pituitary gland cells secrete human growth hormone while others secrete thyroxin or other hormones.

Q: But how exactly do human growth hormone cells "know" what to do?
The question of how DNA directs the development of a human being has fascinated scientists for centuries—and will probably continue to do so for centuries more. There are many questions still left to be answered about genetics, heredity, and cellular biology. For instance, we still don't understand what initially triggers cells to differentiate or precisely how the process takes place.

We do know that the primary function of DNA is to direct the formation of proteins, compounds central to all the processes of life. Proteins control virtually all of the chemical reactions that occur in living matter and make up the bulk of the cells' structures. The composition of the protein—its size and shape—determines how that cell will function. A protein is composed of one or more components called polypeptides, and each polypeptide is a chain of subunits called amino acids. Therefore, the number, type, and order of amino acids in a chain ultimately determine the structure and function of the protein of which the chain is a part. Human growth hormone cells consist of chains of 191 amino acids and it is the order and number of those amino acids that direct the activities of HGH.

You should keep in mind that the process of growth and development never ceases. For the cells of the pituitary to continue to produce HGH, for instance, they must have access to the right amount of amino acids. That's why a proper diet is so essential to the process of growth and development throughout the life cycle.

Q: I'm pregnant with my first baby. When will nutrition start affecting how much she grows?

It already has. The truth is, your own diet—before, during, and after your pregnancy (if you breastfeed)—dramatically influences your baby's growth and development. You and your ob-gyn should work together to create a nutritional plan that will meet your specific needs, but some general rules exist. First, most women need to increase their daily caloric intake by about 300 calories and gain some weight during their pregnancies. (Women of average weight-for-height, for instance, should gain about 25 to 35 pounds.) Furthermore, in order to help your baby grow and develop within the womb and then later in life, you must provide her with enough vitamins and minerals. In general, pregnant women should get 15 to 50 percent more vitamins and minerals each day than women who aren't carrying a child.

One of the most important nutrients for pregnant women is protein, which provides the growth element for body tissues, including the growing baby, the placenta, blood components, and amniotic fluid. Most doctors recommend that you consume at least three 3-ounce servings of protein (lean meats, low-fat cheeses, eggs, beans, and tofu) per day.

Three other nutrients essential to you and your growing baby are calcium, iron, and folic acid. Calcium is necessary for bones to develop and then maintain their rigidity. The body needs iron in order to form hemoglobin, the pigment in red blood cells responsible for transporting oxygen.

Perhaps most important of all to the health of your growing baby is folic acid, a vitamin essential for the process of cell division and the development of healthy tissues. Without it, a fetus runs a much greater risk of being born with spina bifida (an open spine) and anencephaly (absence of a portion of the brain and skull) because it doesn't have an essential raw ingredient it needs to form all of its tissues. Foods rich in folic acid include leafy green vegetables, liver and other organ meats, and eggs. In September 1992, the United States Public Health Service recommended that all women who are capable of becoming pregnant consume 0.4 milligrams of folic acid a day. Once you're pregnant, you should increase to at least twice that amount.

It is also crucial that you avoid all so-called "recreational" drugs, as well as prescription and over-the-counter medications (unless specifically prescribed by your doctor). Any drug—from cocaine to aspirin—can be dangerous to the unborn child, causing a variety of birth defects or physical and intellectual developmental delays. Excessive consumption of alcohol can lead to fetal alcohol syndrome (FAS), characterized by abnormalities of the face, heart, and central nervous system, accompanied by small head size and retardation in growth and mental development. Even moderate drinking may contribute to spontaneous abortion and low birthweight.

Needless to say, then, the process by which new life begins and then develops is vastly complex and depends on several factors—some you can control (such as your diet) and some you cannot (the genetic information you and your husband contribute). The miracle is that, in most cases, the one tiny cell you and your husband formed develops into a whole unique individual person, with all of the organs, tissues, and chemicals necessary for life intact.

Q: I think I understand a little bit about how cells differentiate. But then how do they work together to perform all the body's functions?

At least on a biological basis, you can think of your body as a combination chemistry lab and vast communications center. The endocrine system, which creates chemical messengers called hormones, works in close association with the nervous system, which sends its messages via electrical impulses through nerve fibers. Through these messengers, the neuroendocrine system directs all of the body's activities.

In a way, then, you could also think of the organs and tissues of your body as a symphony orchestra, with the neuroendocrine system acting as conductor. Just as each section of the orchestra plays its own notes, so too does each organ, tissue, and chemical perform a specific function in human physiology. Indeed, each tissue and chemical has its own unique and crucial part to play at exactly the right moment and in direct coordination with one another.

Q: What exactly is a hormone?

The word hormone comes from the Greek word Hormaein, "to set in motion, to spur on, to excite." A hormone is a chemical messenger that takes information from the endocrine gland that secretes it to a specific organ or tissue. When an endocrine gland receives a signal, either from another gland or from nerve cells, it releases hormones to carry instructions to specific cells about temperature changes, hunger, growth needs, or other stimuli.

Humans have about fifty different kinds of hormones, which vary in their structure, action, and response. Hormones travel through the body, either in the bloodstream or in the fluid around the cells, looking for target cells. Once a hormone finds a target cell, it binds with specific protein receptors inside or on the surface of the cell and then specifically changes the cell's activities. The receptors read the hormone's messages and carry out the instruction by either influencing the way the cell's genes express themselves or altering the cell's protein activity.

Hormones vary in their ranges of targets. Some types of hormones can bind with compatible receptors found in many different cells all over the body. Other hormones are more specific, targeting only one or a few tissues. The female sex hormone estrogen, for instance, binds to special estrogen receptor sites in uterine, bone, breast, and other body cells. At the same time, target cells often contain receptors for more than one hormone. The same uterine, breast, and bone cells that accept estrogen, for instance, also contain receptors for progesterone, androgens (male hormones), vitamin D, and other cells.

The receptor then carries out the hormone's instructions by starting one of two cellular processes. It either turns on genes to make proteins in order to produce long-term effects such as growth or sexual maturity, or it alters the activity of existing proteins to produce rapid responses such as a faster heartbeat or higher blood sugar levels.

Q: What hormones directly affect growth and development?

Coordination of growth and development requires a very precise and well-timed means of regulation. Of critical importance are the hormones produced by what is known as the hypothalamus-pituitary axis. These two groups of cells form a unit that exerts control over several endocrine glands, including the thyroid, adrenals, and gonads—as well as a wide range of physiological activities.

What happens is this. The hypothalamus secretes

neurohormones into blood vessels that carry them to the anterior pituitary, a group of cells on the pituitary gland responsible for producing five types of hormones directly related to growth and metabolism. Let's take them one by one.

- **Thyroid-stimulating hormone**. Also known as thyrotropin, this hormone stimulates the thyroid gland, located in the front of the neck at the base of the throat. The thyroid gland releases thyroxin, a hormone that regulates heart rate, force, and output. Thyroxin also helps to regulate the release of HGH, and then works with the hormone to promote skeletal maturation. It also promotes central nervous system growth, stimulates the making of many enzymes, contributes to muscle tone and vigor, and helps the thymus gland produce certain immune system cells.

- **ACTH**. Adrenocorticotropic hormone is responsible for triggering the adrenal glands to release three categories of hormones: mineralocorticoids, glucocorticoids and androgens. Mineralocorticoids help maintain water and salt balance in the body, and thus work to regulate blood pressure. Glucocorticoids are a type of steroid hormone involved in the stress response. When we sense danger, our heart rates speed up, our muscles tense, our breathing rate intensifies: These responses all occur because the glucocorticoids (including norepinephrine and epinephrine) act on the heart, muscles, and other tissues. Androgens are male steroid hormones responsible for secondary sex characteristics. Men produce most of their androgens in the testes, but both women and men produce small amounts of androgens in their adrenal glands.

- **Follicle-stimulating hormone** and **luteinizing hormone**. These so-called gonadotrophs act on the gonads (testes or ovaries) to regulate the production of sex hormones—estrogen and progesterone in women and testosterone in men.

- **Prolactin**. This hormone stimulates milk production by the breasts. The hypothalamus triggers its release after childbirth when estrogen levels are high and in response to the newborn's suckling.

- **Growth hormone**. Also known as somatotropin, growth hormone acts on body cells to promote growth, stimulates the mobilization of lipids from fatty tissue, and promotes secretion of insulinlike growth factors from the liver and other tissue.

As you can see, growth and development depend on far more than the presence of growth hormone. Each of the pituitary hormones stimulates a cascade of physiological events, many of them requiring the presence of more than one hormone. And again, the endocrine system is extremely interdependent. The level of growth hormone, for instance, is intricately linked to the amount of thyroid hormone in the bloodstream and vice versa: Not only does HGH trigger thyroxin release, but through a complex feedback system, the presence of thyroxin also stimulates the release and action of HGH.

Q: Does everyone produce these hormones in the same amount?

Not at all. Each one of us has our own unique chemistry and metabolism. That's what makes it so difficult for physicians to prescribe replacement hormones in the correct doses: No standard dosages of insulin for diabetics or estrogen for post-menopausal women exist, for instance. Instead, doctors and patients work together to adjust the amount and type of medication to fit the patients' physiological needs. It often takes several tries before a patient receives just the right dose.

That's one reason many experts are concerned that anti-aging hormones like melatonin and DHEA may be too accessible. The fact is, each and every one of us has a different level of need for these powerful chemicals—and some of us may not need any at all. Nevertheless, manufacturers market them in standard doses and leave it to consumers to monitor their short-term effects and side effects. And the long-term consequences of anti-aging hormone replacement remain unknown. Even the use of estrogen, which has been in wide-spread use for almost thirty years, remains controversial, with many studies indicating an increased risk of breast and endometrial cancer associated with both estrogen replacement after menopause and the use of oral contraceptives.

Q: Does fetal growth depend on human growth hormone?

Not precisely. It isn't until after birth that a baby's pituitary gland develops enough to secrete its own

HGH. Instead, a different but chemically related chemical called human placental lactogen helps the fetus and newborn grow. Human placental lactogen, also called placenta protein, is found in the mother's blood and in the placenta, the blood-rich structure in the uterus through which the fetus takes in oxygen and other nutrients.

Q: When does human growth hormone start to influence growth?

After the first few weeks of life, a child's pituitary gland begins to secrete HGH and other hormones and chemicals that direct growth. In fact, the first few years of life represent the largest growth period of a child's life until adolescence. In just four to five months, a healthy baby almost doubles his birth weight, and by one year, he triples it. A baby grows about an inch every month or so during his first year, while his organs and tissues rapidly grow and develop.

After the first year, the rate of physical growth slows down a bit, with a child growing about 1.5 to 2 inches per year. Usually by about two or three, a child develops an established pattern of growth relative to his peers, a pattern he maintains until puberty—what parents know as the "growth curve." A child's growth is considered "normal" if it remains within this growth curve, and follows the established pattern. For example, doctors consider a child who falls within the middle range of the curve, and remains in the middle range from year to year, to be progressing normally. When a child falls outside the growth curve, or when the growth pattern suddenly changes, there may be reason for concern.

Q: What happens to growth during puberty?
Interestingly enough, the word *puberty* is derived
from the Latin *pubescere*, which means to be covered with hair, an apt description of the growth of
pubic and body hair that comes with adolescence.
But puberty brings with it far more intense and
complex changes, with the body reaching adult
stature and physical maturation, the reproductive
organs maturing to become fully functional, and
psychological changes occurring to adapt to these
physical and sexual developments.

During this period, the body grows at a faster
pace and the rate of skeletal maturation increases.
This growth spurt tends to occur earlier in girls
than in boys, though the later onset in boys results
in a longer growth period. In boys and girls, both
leg and trunk growth contribute to the overall increase in height. At some point during adolescence—and the age varies from child to child—the
longitudinal growth of bones stops when the epiphyseal plates, the rounded heads of long bones,
become solid and stop producing new cells. After
this point, we can't grow any taller nor can the
length of our arms and legs increase.

**Q: How does a boy's musculoskeletal system
     differ from a girl's after puberty?**
Before puberty, boys and girls have a very similar
body shape and composition. In the very early
stages of puberty, lean body mass increases dramatically, building muscles in both girls and boys.
Once girls start to produce lots of estrogen (just before and following their first period) and boys start
to produce testosterone, body composition begins
to change. Boys continue to increase muscle mass

throughout puberty, to reach values much higher than those in girls. While boys tend to lose body fat during adolescence, girls tend to gain it, to values almost twice those found in boys.

The reason for these differences in body type can also be traced to hormones: Male hormones (androgens) are much more effective muscle builders than female hormones (estrogens). That's why anabolic steroids—synthetic versions of testosterone— are used by athletes to increase muscle mass, a subject to which we'll return in Chapter Four.

Q: How common are growth problems?

According to the Human Growth Foundation, an estimated five hundred thousand people in the United States currently suffer from various growth problems, the vast majority of them children.

- **Growth hormone deficiency syndrome**. Doctors consider this condition "classic" HGH deficiency, and it occurs when the pituitary gland does not secrete sufficient levels of HGH. Although growth hormone deficiency syndrome can occur in adults, physicians have yet to come up with strict criteria for diagnosis.

- **Hypopituitarism**. Hypopituitarism is a medical condition in which the pituitary fails to release its hormones, usually because of a tumor, infection, or surgery. In this case, sexual development and sugar metabolism often also suffer, since the pituitary fails to release hormones that regulate these activities in addition to HGH.

- **Turner's syndrome**. Due to a genetic defect that affects only females, Turner's syndrome results not only in short stature but a host of other hormonally–related problems, including those related to sexual development. Mental retardation may also occur. Although the pituitary gland secretes normal amounts of growth hormone, girls suffering from Turner's syndrome attain heights of only 4'8" or less.

- **Chronic renal insufficiency**. This disorder of the kidneys primarily affects children, causing both kidney- and growth-related problems. Now that dialysis and kidney transplants save lives that would otherwise have been lost to kidney disease in childhood, HGH is necessary to help these children grow normally.

Q: Does HGH deficiency cause someone to be a dwarf?

There are two different types of disorders classified as dwarfism: disproportionate short stature and proportionate short stature. Disproportionate short stature, also known as achondroplasia, is a genetic disorder, characterized by a height of less than about 4'10", a prominent head, and a waddling gait. People with achondroplasia produce enough HGH, but their genetic defect prevents the cartilage at the ends of bones from growing normally. Lack of HGH, on the other hand, is the cause of most cases of proportionate short stature, characterized by height less than 4'10" but with normal-sized head, trunk, and limbs.

Q: Is dwarfism hereditary?

Since achondroplasia is a dominant disorder, a man or woman with the disorder runs a 50 percent risk of bearing a child with the same condition. It turns out, though, that the vast majority of achondroplastics have parents of normal height, which means that this condition mainly occurs "out of the blue." The same pattern of inheritance appears to be involved in proportionate short stature as well.

Q: Does treatment with recombinant HGH (rHGH) help children with growth hormone deficiency?

Scientists estimate that about ten to fifteen thousand American children are short because of a growth hormone deficiency, and for these children rHGH is an essential part of treatment. With rHGH, their bones and muscles obtain the chemicals they need to grow and develop normally. Without question, those who respond well to rHGH will be taller as adults than they would have been without treatment. HGH also affects body composition by limiting body fat and influences metabolism by helping the body to properly process cholesterol and glucose. By doing so, rHGH helps prevent these children from developing serious illnesses later in life.

Although the use of rHGH in children with other types of growth problems remains controversial, some doctors believe that stimulating the growth of tissue by providing rHGH could help other children reach their height potential. Some studies show that rHGH therapy triggers growth of about 4 inches a year. Other studies indicate that the treatment merely speeds up the rate of growth, but does not affect total height.

Q: Are there side effects associated with rHGH therapy in children?

Because doctors have been prescribing rHGH for such a short time, they haven't had a chance to assess long-term side effects. Unproved but suspected side effects in children include leukemia, exaggeration of scoliosis (an abnormal curvature of the spine), swelling, allergy, and impaired glucose tolerance. What's more, there is some evidence that too much IGF-1, a growth factor whose secretion from the liver is stimulated by the presence of HGH, could trigger the development of breast and/or colon cancer later in life.

The National Institute of Child Health and Human Development cautions doctors to wait for more definitive studies to be conducted before prescribing rHGH to children with rHGH deficiency, Turner's syndrome, or another diagnosed growth hormone deficiency. In those children, the risk of side effects and complications may well outweigh the benefits.

Furthermore, doctors and parents alike should keep a close watch on the psychological development of the child on rHGH therapy. For the most part, self-esteem and body image improve considerably as a child begins to grow, and the child is likely to gain self-confidence and a sense of optimism about the future. However, some children expect too much from the treatment, assuming they will sprout up overnight or become the tallest kid in class; when their expectations aren't met, they feel disappointed, depressed, or even guilty. Others regret losing the special niche they have in the family or at school when they begin to grow, and are uncomfortable with their growing bodies.

A pediatric endocrinologist—one who understands both the physical and emotional effects of rHGH therapy—should closely monitor any child taking rHGH, which means an appointment every three or four months with periodic laboratory and x-ray evaluations and dose changes. Treatment continues until the longitudinal growth of the bones stops when the epiphyseal plates—the round ends of the bones that produce new cells during childhood—close.

Q: How should my physician evaluate my daughter for rHGH therapy?

Even endocrinologists find it extremely difficult to diagnose HGH deficiency with accuracy, so your first step is to find a qualified professional, one who has a great deal of experience in this area of medicine. Your pediatrician should be happy to recommend someone for you to see.

Why is HGH so difficult to diagnose? As we discussed in Chapter One, the amount of HGH itself is hard to measure since the pituitary releases it in short bursts and the liver almost immediately metabolizes it. Although the physician will attempt to measure it by performing a variety of x-rays and laboratory tests, including those that test for IGF-1 levels, he'll also conduct a thorough examination and family history.

The doctor will ask how tall other members of the immediate family are, and if any family members have been diagnosed with a disorder associated with short stature. He'll also evaluate your child's height in relation to other children her age: Is she at the bottom of the normal growth curve (the lowest 3 percent for height)? Does she show signs of delayed skeletal development and maturation? Is she suffering from low self-esteem or has she failed to progress socially because of her height? Perhaps most importantly, the doctor will judge her speed of growth: Is the speed of growth consistent and within a normal range? Is the child's growth rate slowing down when it should be steady or increasing? Learning the answers to these questions and making a decision about the efficacy of rHGH therapy can take months of careful height evaluations performed by a doctor on a regular basis.

Q: How many kids who are short, but appear to produce enough natural HGH, receive rHGH treatment?
According to a survey conducted by researchers at the Department of Pediatrics at Case Western University Medical School and published in the August 17, 1996 issue of the *Journal of the American Medical Association*, approximately 58 percent of children undergoing rHGH therapy have classic growth hormone deficiency, while 42 percent were short for other reasons—genetic or medical. Among the conditions listed by the responding physicians were Turner's syndrome, chronic renal insufficiency, and familial or unknown causes of short stature.

Q: Does rHGH help children without a clinical HGH deficiency?
The authors of the study, Leona Cuttler, M.D. and colleagues from the Case Western University School of Medicine, cite research that suggests that short-term HGH therapy does help to increase the

rate of growth in a majority of children who take it—whether or not they produce sufficient HGH. What we don't yet know, however, is if rHGH does anything more than speed growth in these children while having no effect on their ultimate height. Considering that there are serious side effects, then, most physicians, including those who conducted this study, feel that doctors and parents should seriously consider whether or not the risks of therapy outweigh the as yet inconclusive benefits.

Q: Why should a short child, but one without a clinical HGH deficiency, receive rHGH treatment?

Many pediatricians believe that short stature may have a negative emotional impact on both children and adults. About 84 percent of physicians who responded to the survey conducted by the Case Western researchers thought that short stature usually or often impairs emotional well being in children between the third and fifth percentiles for height. This number rose to 97 percent for children below the third percentile in height. In an editorial that accompanied the study in the *Journal of the American Medical Association*, Barry Bercu, M.D., writes that "Parental pressure to mitigate short stature in their children is driven by a cultural 'heightism' that permeates American society. Taller college graduates make more money, and 80 percent of U.S. presidents have been the taller candidate. Unfortunately, there are no definitive long-term controlled studies describing psychological outcome, but several psychological studies indicate that short stature per se does not result in negative psychological adaptation."

Nevertheless, doctors continue to prescribe this potentially risky medication to children without a clinical HGH deficiency—especially if their parents want the treatment. All other factors being equal, physicians were more likely to prescribe growth hormone if the children's families were strongly in favor of the treatment.

Q: Are there adults with growth hormone deficiencies?

No one knows exactly how many adults have a clinical HGH deficiency. Some adults with HGH deficiency have it because their doctors discontinued HGH therapy when longitudinal bone growth ceased during puberty. They did so both because cadaver pituitary HGH was scarce and the benefits of HGH after puberty were unknown. Acquired HGH deficiency results from the destruction of normal pituitary and/or hypothalmic tissue, usually from a tumor or secondary to surgical and/or radiation therapy.

Q: How does having an HGH deficiency affect an adult's health?

That depends on how severe the problem is and when it occurred. People with HGH deficiency often complain of fatigue, lack of concentration, memory difficulties, and irritability in addition to weight gain, increased fat mass, and decreased lean body mass. Depression and lethargy also seem to evolve as HGH deficiency continues.

Women of short stature (under 4'10") due to HGH deficiency in childhood may have difficulties during pregnancy. Some pregnant women with dwarfism tend to have respiratory problems because the uterus presses against the chest cavity as the fetus grows. Others may experience nerve root compression, which results in numbness and tingling of the lower limbs. Because of their small pelvic structure, most small-statured women must deliver their children by caesarean section.

In one recent study, people with pituitary tumor-related HGH deficiency suffered a 7 percent increase in fatty tissue (largely in their abdomens) while their lean body mass decreased to a similar degree. At the same time, their blood lipid levels—cholesterol and triglycerides—increased. Their muscle mass and strength both diminished, and in the heart, these changes result in a reduced left ventricular mass and decreased cardiac output.

In a study published in a 1991 issue of the *Journal of Applied Physiology*, exercise capacity decreased by 20 to 25 percent compared to normal controls. Yet another study shows how HGH affects bone density in adults: In a 1993 issue of the *Journal of Clinical Endocrinology and Metabolism*, researchers revealed that cortical and spinal bone density were 2.8 and 1.5 standard deviations below the mean for age and sex matched controls. The results of these studies, and others, show that the need for HGH never ceases, even after longitudinal growth ends in puberty. What we still don't know for sure is how effective rHGH therapy is in adults.

Q: Can children or adults have too much HGH?
As mentioned in Chapter One, too much HGH produces gigantism in prepubescent children and acromegaly in adults. Gigantism is extremely rare, and is associated with enlarged soft tissues and late

skeletal growth plate closure. Children with gigantism may reach a height of up to 7 to 8 feet. As with dwarfism, a variety of genetic and hormonal causes not related to HGH levels are the most common causes of gigantism.

In adults, excessive growth hormone results in a condition called acromegaly. In this condition, excessive levels of HGH cause soft tissue and cartilage to grow abnormally, including that of the hands, feet, larynx, and sinus. Facial features become distorted due to additional growth of the jaw, brow, and nose. Women with acromegaly run a higher risk of developing breast cancer, and both men and women with the disorder are at greater risk for colon cancer.

Q: We've talked a lot about height. What about weight? What determines a person's body type?

As is true for height, one's body type depends primarily on two factors: genetics and nutrition. It appears we are all born with certain body type traits: We grow only so tall, our muscle fibers are only so long and lithe, and our bodies use energy in certain specific ways, ways that we inherit from our parents. However, as is true for so much about human physiology, personal habits and environmental influences may outweigh the genetic component in weight-related matters.

Q: Why is being overweight such a common problem today?

The regulation of eating behavior and metabolism is one of the least understood aspects of human physiology, mainly because so much more than mere hunger and satiety are involved. If we ate only when we were hungry, and only consumed what our bodies needed to survive, the diet-related problems of diabetes, cardiovascular disease, and other disorders would be severely diminished. However, there are so many social and cultural factors involved in eating that treating it as a simple biological process is impossible. In Chapter Nine, we'll discuss some of those reasons, and how you can work to counteract them to lose and then maintain a healthy weight.

Q: What exactly is metabolism?
Food provides the body with the energy it needs to function. Metabolism is the process by which the body breaks down food, processes it into glucose and other substances that act as fuel, and then uses that fuel as energy. The rate of your metabolism depends on a variety of genetic and lifestyle factors.

Q: How does food become fuel?
The food we eat is converted into fuel through the process of digestion. Digestion is both a mechanical and chemical process that begins in the mouth, where the jaws and teeth break down food and mix it with saliva. Saliva contains amylase, the first of many enzymes involved in digestion. After you chew and swallow food, it moves into the pharynx and then into the esophagus, a hollow, foot-long, muscular tube that carries food to the stomach. The stomach is a pear-shaped muscular organ situated mostly on the left-hand side of the body, below the lower ribs. Contraction of the three layers of stomach muscle mixes the food with digestive enzymes and hydrochloric acid that complete the digestion

of food, breaking it down into its basic parts.

The food we eat is made up of carbohydrates, proteins, fats, vitamins, minerals, and water. When these substances pass through the stomach and small intestine, enzymes break down the long molecules of each group: carbohydrates break down into glucose, proteins into amino acids, and fats into fatty acids. Once this process is complete, these components pass through the wall of the small intestine and into the bloodstream for the body to use as energy and to aid in growth processes.

Glucose circulates in the bloodstream, coming into contact with all cells. It even passes through the kidneys, which rescue it from waste products and return it to circulation. Glucose also comes into contact with the pancreas, a gland situated behind the stomach. This endocrine gland is crucial to the way the body uses glucose as energy. It produces two hormones: glucagon, which raises the level of glucose in the blood; and insulin, which triggers cells of the body to use glucose as energy. Liver and muscle tissue store as much excess glucose as possible, keeping it readily accessible (as a substance called glycogen) until the body needs it for a quick supply of energy. The problem is that the liver and muscles have limited storage space. Any glucose left over is stored as fat.

Amino acids, derived from the digestion of protein, help build and repair muscle and bone (by working with HGH to synthesize proteins) as well as produce enzymes. The body uses about 50 percent of amino acids as energy by converting amino acids into glucose in the liver. Amino acids, like glucose, can also be changed to fat for storage until the body needs them.

Fatty acids, derived from the digestion of dietary fat, are stored in adipose tissue (fat) for later use. The body only uses fatty acids for energy when other fuels are in short supply, insulin level in the bloodstream falls, and/or the presence of HGH in the bloodstream increases. The reduced level of insulin and higher HGH levels promote the removal of fat from the storage depots and help its entry into the circulation.

Q: This sounds like an efficient process. Why then do we gain weight?
The simple truth is that we gain weight whenever we provide more food to the body than it needs to fuel its activities. As you see, every component of food (except vitamins, minerals, and water) will be stored as fat unless the body uses it quickly for energy. The trick is, each individual uses energy just a little differently from anyone else. That's why people tend to blame their "metabolism" for their body type. We've all witnessed it: Many thin people seem to be able to eat as much food as they like without gaining weight, while many heavy people seem to store food as fat as soon as they consume it.

Q: Is metabolism something you inherit?
To a certain degree, yes. It seems clear that a genetic component to body type exists. Only about 7 percent of children with normal weight parents grow up to be fat, for example, while 40 percent of children with one fat parent grow up to have a weight problem. If both parents are fat, the odds go up to 80 percent. A study reported in the *New England Journal of Medicine* in the mid-1980s evalu-

ated 540 middle-aged adults who had been adopted as children. Their body composition (fat to muscle ratio) bore little relation to that of their adoptive parents. Instead, the daughters tended strongly to follow their biological parents' patterns, particularly of their mothers. The sons, however, showed no such relationship to either set of parents. As you can see, genetics and environment must work together when it comes to weight management.

Of course, parents pass on more than genes to their children. They also pass on eating and exercise habits that impact on the way the body uses energy. As we'll discuss at length in Chapter Four and Chapter Nine, exercise speeds up metabolism by using up glucose quickly and by building muscle, which requires more energy to maintain than fat.

Q: But is there a "fat gene?"

Again, there are probably several genes that help to regulate metabolism and create body type, and scientists continue to search to complete the genetic puzzle. In late 1994, scientists made perhaps their biggest stride when they discovered—and then cloned—a mouse gene that has a role in telling the body to plump up or slim down. They called the gene Ob, for obese. Later, they discovered that the gene is responsible for creating a hormone called leptin. People who inherit the Ob gene do not use this hormone properly and thus tend to gain weight.

Q: How does leptin work?

In a complex feedback loop, fat cells release signals in the form of the hormone leptin, which travels in the bloodstream until it reaches leptin receptors in the hypothalamus. If the level is high, the hypothalamus passes on a signal indicating that the body is storing too much fat. The hypothalamus then sends out instructions (again in the form of hormones) to various parts of the brain and body to reduce appetite or to increase the rate at which fat is burned.

If this system works properly, the body maintains ideal weight. However, a variety of things can go wrong: The gene that makes the signaling molecule, leptin, may fail to make it properly, so the hypothalamus never receives messages about how much extra fat the body carries. Or the hypothalamus may fail to trigger the proper response to the message once it receives it. Researchers continue to trace these complex neuroendocrinological processes in order to find ways to more successfully treat and prevent obesity.

Q: Will taking rHGH help me to lose weight?
That's hard to say. It is true that HGH plays a significant role in metabolism and body composition. It promotes fat loss because it triggers the release of fat cells from adipose tissue for the body to use as energy, instead of using the more readily available blood glucose or even glycogen stores in the liver and muscle. It also helps to build muscle, which requires more calories and energy to maintain than does fat. The combination of these two actions helps create a lean, healthy body.

The best evidence of HGH's dramatic impact is a comparison between the body composition of a twenty-year-old, who has high levels of HGH, and that of a seventy-year-old, whose levels of HGH are about 20 percent of his younger counterpart. Approximately 80 percent of a young adult's body consists of lean body mass (muscles, organs, and bone); the remaining 20 percent is made up of fatty adipose tissue. But after age thirty, the muscles begin to atrophy, the skin thins out, and lean body mass (including the integrity of vital organs) is replaced by adipose tissue at an average rate of 5 percent a decade. Depending on diet and exercise habits, the body composition of a seventy-year-old will be roughly 60 percent lean body mass and 40 percent adipose tissue.

However, most experts insist that diet and exercise remain the most important ingredients for healthy, permanent weight loss and weight maintenance. No matter how effective hormones like HGH and DHEA might be in stimulating your metabolism, you'll still gain weight if you eat too much and/or exercise too little.

Q: Does the body ever go through another change as radical as the one that occurs in adolescence?

Beginning at about age fifty, men and women begin to lose their sex hormones. In women, this loss is more sudden and complete. Called menopause, this stage of life represents the loss of fertility as the ovaries, the woman's primary sex organs, stop producing estrogen and progesterone. Whether there is a male equivalent of menopause remains to be confirmed: men remain fertile well into their eighties but tend to be less sexually active and their androgen levels decline with increasing age. In Chapter Three, we discuss the changes that come with age—and how hormones like HGH can help alter them—in further depth.

THREE

The Biology of Aging

Q: Why do we grow old?
This is a question that can be answered in two ways. In a larger sense, we age and finally die in order to make room on this planet for the next generation of human beings to live and thrive. Should the mortality rate fall—if fewer people die than are born—the Earth could become pathologically overpopulated (something some scientists believe has already occurred), which would strain natural resources and thus put the entire ecosystem and all its life forms at risk.

On the molecular level, the same general theory holds true. When certain individual cells within an organism wither and perish, new cells replace them and function with vitality and health. As mentioned in Chapter Two, each cell contains information about its life cycle encoded in its nucleus. Normally, cells mature, reproduce, and then, in an action called "apoptosis," self-destruct. A disruption of this process may compromise the overall health of the organism. An example of such a disruption in a cell's life cycle is cancer. For reasons not fully understood, cancer cells no longer have, or no longer heed, correct genetic instructions about reproduction and death. They grow indiscriminately, using up nutrients meant to nourish healthy cells, which then die before their time. If left unchecked, cancer cells grow and spread until the organism as a whole is unable to function.

Biologically speaking then, aging and death are essential and quite natural aspects of life. And yet, because of our very "humanness"—the deep emotional and intellectual ties that bind us so intimately to earth and to each other—we continue to look for ways to circumvent this seemingly inevitable process. And the more we learn about the human body and its aging cycle on a molecular level, the more possible such a goal appears to become. In fact, a new science, called biogerontology, has been developed to study just these issues.

Q: What causes the body to age?
In essence, aging and death occur when cells within the body die or malfunction and then are neither repaired nor replaced. Eventually, the system or organ affected by cell death or mutation will no longer be able to operate properly. And, because human physiology is so interdependent, disruption in one part of the body often has widespread effects.

This apparently inevitable loss of healthy cells and the chain reaction of damage that often results lie at the heart of the most commonly fatal diseases of aging—heart disease and cancer—as well as age-related chronic conditions such as arthritis and senile dementia. We think that one reason that we lose HGH as we age, for instance, is that the pituitary gland begins to lose its HGH-secreting cells. Ironically, if HGH continued to flow at youthful levels, it could help prevent or at least delay tissue damage like this. However, until medical science can find a way to halt this decline, our bodies are designed to fail in one way or another as we age.

Q: Do researchers know what sets this process in motion?

Exactly how and why cell death occurs remains the subject of intense interest and study. In fact, gerontologist Zhores Medvedev identified and reviewed more than three hundred different theories of aging in an article he wrote in 1990 for *Biological Review*—and all of them appear to have some merit. For the purposes of this discussion, we'll focus on four of the most generally accepted hypotheses.

- **Programmed cell death**. Some scientists ascribe to the theory that every organism has an innate, species-specific life span. That's why most dogs live for about twelve years, fruit flies approximately twelve months, and human beings about seventy-five years, to cite just a few examples. According to this theory, death, like all other physiologic activities, is genetically programmed into an organism's every cell. Exactly which and how many genes are directly involved in regulating human aging is still unknown, but some geneticists estimate the number to be about two hundred.

- **Free radical damage**. Another theory about aging centers on substances called free radicals. A free radical is a molecule that possesses an unpaired electron, one that is constantly attempting to become whole by robbing components from healthy cells. Unfortunately, this process—also called oxidative stress—damages healthy cells, sometimes beyond repair. Scientists think that the body becomes more vulnerable to free radicals with advancing age: Not only do we produce more free radicals, we are also less able to defend ourselves against them. Free radicals then batter our proteins, cell membranes, and DNA. They clog the wall of our arteries, kill brain cells, stiffen and deplete our muscles, and throw our immune systems out of kilter. The damage done to enzymes, cell membranes, proteins, and DNA by these harmful substances may lead to the development of several age-related diseases, such as atherosclerosis, cancer, Alzheimer's disease, and others.

• **DNA damage**. Some scientists postulate that the symptoms and side effects of the aging process could occur as a result of a series of mistakes made in the genetic blueprint encoded in every cell's DNA. As a cell ages, it becomes less accurate in its reproductive process for a variety of reasons, including an age-related increased vulnerability to free radical damage as well as a decreased ability to repair itself. When genetic errors accumulate, the tissue the cells comprise becomes damaged, unable to function properly. Depending on what part of the body is involved, this degeneration of cells could lead to any of a number of age-related illnesses. DNA damage to cells of the pituitary, for instance, could severely compromise the release of HGH.

- **Hormonal disruption and failure**. Another theory about the cause of aging centers on the endocrine system. Indeed, some biogerontologists believe there is an aging clock somewhere in the brain—probably in the hypothalamus—that directly influences the slowdown in hormonal production. As we've discussed, the amount and type of hormones our body secretes dramatically changes as we age. This alteration in hormone secretion eventually results in a bodywide imbalance that affects our reproductive ability, the health of our tissues (including the muscles, bones, and organs), and the functioning of our metabolism and immune system. We therefore become ever more susceptible to disease and disability.

Q: So which theory explains why older people tend to get sick more often, and with more chronic and ultimately fatal illnesses, than young people?

It is likely that aspects of each of these theories, and others, affect the aging process. It is clear, for instance, that men and women have a rather fixed amount of time to live (no one has survived beyond the age of 120 and most people die at about seventy-five)—a duration that genes program in some way. In other words, every cell in the body knows when to perform apoptosis, or programmed cell death.

At the same time, scientists link free radical damage to the development of several age-related diseases. Osteoarthritis (the wearing away of cartilage), for instance, may be due at least in part to the destruction of healthy cartilage by free radicals. Another example is Parkinson's disease, which involves the gradual but inexorable death of certain brain cells. Parkinson's disease tends to develop in old age after a lifetime of exposure to free radicals and the chemical warfare they wage on the body and the brain.

More and more, however, scientists are considering the possibility that the endocrine system, specifically the hypothalamus and the endocrine glands it stimulates, might lie at the center of the mystery. Its effect on the body is so widespread, and its involvement in growth and maturation throughout life so intimate, that many people believe its role in the aging process is equally pivotal.

Q: How does the amount of HGH change as we age?

As discussed in Chapter One, HGH levels are high during childhood, peak in adolescence, remain steady until the age of about thirty, then slowly decline until they reach very low levels after the age of about sixty. The same pattern holds for the insulinlike growth factors, IGF-1 and IGF-2, whose production is triggered by HGH and that work together with HGH to maintain growth and metabolism.

Q: How does the loss of HGH affect the aging process?

Virtually all of the complications of aging stem from the body's inability to replace cells as it loses them to programmed cell death, free radical damage, or DNA mutations. To replace the cells—in other words to build new cells—the body must be able to synthesize protein which makes up the bulk of a cell's composition. Indeed, one of the characteristic signs of aging is a decline in protein synthesis in virtually every cell of the body. The process by which the cells synthesize protein depends to a large degree on HGH. When HGH is insufficient, neither repair nor replacement can occur. In addition, HGH's role in stimulating the production of immune system cells and maintaining immune system organs may also prove to be a factor in the aging process, since without a healthy immune system, we are vulnerable to life-threatening infections. We'll discuss this aspect in Chapter Seven.

Q: How do we know that replacing HGH helps reverse these problems?

The most famous study about the relationship between rHGH and aging was performed by Daniel Rudman, M.D. at the Medical College of Wisconsin in Milwaukee and published in the July 5, 1990 issue of the *New England Journal of Medicine*. He and his colleagues studied twenty-one healthy men from sixty-one to eighty-one years old who had low levels of plasma IGF-1, a sure indication that their HGH levels were also low. During the treatment period, twelve men received approximately .03 mg of rHGH per kilogram of body weight subcutaneously three times a week while nine men received no treatment. Every month, the researchers measured lean body mass, body fat, skin thickness, and bone density.

The results proved very exciting. After six months, the men in the first group experienced an increase in IGF-1 levels to a youthful range of 500 to 1,500 U per liter from 350 U per liter. Their lean body mass increased by 8.8 percent, their body fat decreased by 14.4 percent, their skin thickness by 7.1 percent, and their vertebral bone density by 1.6 percent. The men in the second group, on the other hand, experienced no significant change in lean body mass, fat distribution, skin thickness, or bone density.

In this study, all of the men remained healthy and none had any changes in the results of differential blood count, urinalysis, or blood chemistry profile. None suffered any expected side effects such as swelling, carpal tunnel syndrome, or allergic reactions to the HGH. However, they did experience small but manageable increases in systolic blood pressure and blood sugar.

Q: Do other studies confirm Rudman's findings?

The results have been mixed, and most experts agree that we need much more research before we fully understand the effects of replacing HGH in late life. An animal study performed at North Dakota State University in 1996 appears to support the enthusiasm engendered by the Rudman study. Researchers gave one group of mice injections of saline (salt and water) solutions while another group received injections of growth hormone. After thirteen weeks of treatment, 39 percent of saline-treated mice were still alive, while 93 percent of the HGH-treated mice survived. Injections were then stopped for six weeks. During this period, all of the remaining saline-treated mice died of old age, while only one in twenty of the HGH-treated mice died. The researchers then resumed injections of growth hormone on the remaining nineteen mice for another six weeks, after which time eighteen mice survived.

Q: What about human studies?
In 1996, Dr. Maxine Papadakis performed a study of rHGH's anti-aging effects at the University of California at San Francisco. She and her colleagues studied fifty-two men, aged seventy and older, all with levels of growth hormone normal for their age—which is about 10 percent of the typical level of forty-year-olds. Half of the subjects took injections of growth hormone three times a week, while half took placebo shots containing no hormone. The study was double-blinded: Neither patients nor researchers knew who received hormones and who received placebos until after the study was completed.

At the end of six months, the researchers found changes in body composition similar to those reported by Rudman: patients taking the hormone experienced a 4 percent increase in lean body mass and a 13 percent decrease in fat, with no change in weight.

However, when the researchers performed further tests to evaluate muscle strength, endurance, and mental ability, they found no difference between the two groups. The group that took rHGH injections were no stronger or more alert than those who received placebos. And those who took the hormones suffered side effects, including swelling in their ankles and lower legs and joint stiffness and pain, especially in their hands. About 26 percent of the patients felt these side effects so severely during the study, they asked the researchers to lower the dose of hormone.

Q: What could account for such extreme differences in results between the Rudman and Papadakis studies?

Dr. Papadakis believes that the main reason for the disparity between the two studies concerns the methods used to conduct them. In Rudman's study, everyone knew they were taking a hormone meant to improve their health and sense of well-being. In an April 15, 1996 article in the *New York Times*, Rudman's widow and colleague, Inge Rudman, remarked, "I think those men were expecting to feel better. And they did. They became euphoric. Sometimes it's the psychological effect of expecting to feel better."

Nevertheless, the fact that a measurable increase in bone and lean muscle mass occurred in both studies points to the importance of HGH in maintaining body tissue. What scientists must determine now is just what that increase in mass means to the overall health of the body, and if replacing HGH leads to increased longevity or well-being.

Q: What other hormones affect the aging process?

As discussed at the beginning of the chapter, many scientists studying how and why we age believe that the ebbs and flows of the endocrine system represent the keys to the process by which we live, age, and finally die. Indeed, we now know that we lose several hormones that perform essential functions as we get older, including the following.

- **Estrogen**. Women begin to lose estrogen after the age of thirty-five and pass through menopause at the average age of fifty-two. Although the loss of estrogen most directly affects the reproductive system, it also has an impact on a number of other physiological processes. Estrogen plays a role in maintaining the health of the skin, protecting the brain from degenerative diseases like Alzheimer's disease and mood disorders like depression, to say nothing of defending heart and bone tissue from free radical attack.

- **Testosterone**. Although the age-related health effects appear to be less severe than those triggered by the loss of estrogen, men, too, lose their primary male sex hormone, testosterone, as they age. By about age forty-five, testosterone levels begin to decline, and from that point on, testosterone production drops slowly but steadily with each passing year. Testosterone deficiency causes a host of problems, including loss of libido, depression and other mood disorders, fatigue, muscle weakness, and—just like women who lose estrogen—osteoporosis. Furthermore, some biogerontologists believe that women, too, lose androgens as they get older, and that the loss of these male hormones triggers a decrease in vitality and libido in the female sex as well.

- **DHEA**. This adrenal hormone plays a role in the formation of testosterone and estrogen. Production of DHEA is highest during fetal development, rises during adolescence, reaches maximum levels at about age thirty, and declines after that. On average, DHEA levels in people over age sixty-five are about 15 percent of those found in twenty-year-olds. DHEA helps to stimulate immune system function, improve glucose metabolism, and suppress the development of many cancers. Without it, then, the older body appears to be far more vulnerable to these conditions.

- **Melatonin**. Produced by the pineal gland, melatonin levels rise and fall in a pattern similar to that of HGH and DHEA: It is highest during adolescence, remains steady until about thirty or thirty-five, then slowly begins to decline as the pineal gland begins to shrink. Some scientists believe that melatonin's presence helps to coordinate virtually all of the endocrine system's activities by acting as the body's timekeeper. Indeed, the pineal gland keeps the body in sync with the most constant environmental cue we have: the light-dark cycle. Through its release or suppression of melatonin, the pineal gland announces to the rest of the body that it is dawn or dusk, time for the body to be awake and alert, or time to prepare for bed and rejuvenating sleep.

This crucial signal sets complex processes into motion, a cycle designed to remain regular and balanced. Called a circadian rhythm, this cycle may become unbalanced as we age due to the lack of melatonin and other factors, and thus may foster the aging process itself.

Q: What is circadian rhythm?

From the Latin "circa" (about) "diem" (a day), the circadian rhythm is the twenty-four-hour cycle of light/dark, wakefulness/sleep to which most human physiologic activities are set. At regular intervals each day, the body tends to become hungry, tired, listless, energized. Body temperature, heartbeat, blood pressure, hormone levels, and urine production rise and fall in a relatively predictable, rhythmic pattern—a pattern initiated and governed by exposure to sunlight and darkness.

In the modern world, saturated with artificial light and loaded with constant stimulation, most of us attempt to rely completely on external signals to put our biological clocks in motion. We are not awakened at dawn by the sun, but earlier or later by an alarm clock. At night, the flickering of the television monitor keeps us up and stimulated long after natural nightfall might have triggered our bodies to prepare for sleep.

Nevertheless, we retain a stubborn attachment to what seems a rather primitive code of behavior established by our most constant and eternal signal: the rise and fall of the sun. When we deviate from the natural pattern, by traveling across time zones or working the night shift, we often become physically and mentally disoriented until we reestablish our connection to it. The widespread importance of circadian rhythms to every aspect of our well-being has caught the attention of physicians, biogerontologists, pharmaceutical companies, and millions of health-conscious men and women.

Q: Isn't melatonin responsible for sleep/wake patterns?

Nicknamed "the chemical expression of darkness," melatonin—like HGH—is produced almost exclusively at night or in a light-free environment. The high concentration of nocturnal melatonin led scientists to conclude that the production of this hormone signals to the rest of the body that it is time to sleep. Indeed, melatonin supplements have been used for decades to treat sleep-related problems, such as insomnia, sleep apnea, and jet lag. In the morning, when we perceive that it is light, melatonin secretion ceases, which stimulates the production of other hormones and hence other body activities to begin.

This orderly daily rhythm of hormone secretion is of prime importance to our physical, intellectual, and emotional health. The blood level of melatonin appears to trigger the adrenal glands and gonads to increase or suppress the secretion of male and female sex hormones, as well as the stress hormones, norepinephrine, epinephrine, and cortisol. Its presence or absence also affects the production of pituitary gland hormones, including HGH, which, as you know, is produced largely at night, when melatonin levels are high as well. The presence or absence of HGH itself triggers certain hormonal changes, including the release or suppression of growth factors, IGF-1 and IGF-2, which perform a wide range of bodily activities.

Q: How does the brain know when it is light or dark?

Once again, a crucial activity can be traced to the hypothalamus: The same set of brain cells responsible for triggering the release of HGH and other hormones from the pituitary also acts as the body's timekeeper. The hypothalamus contains two clusters of cells called the suprachiasmatic nuclei (SCN). The SCN, made up of more than eight thousand brain cells each, act as the body's circadian pacemaker. In mammals, the SCN appear to get their information from photoreceptors in the retina, which transmit signals about light and dark through the optic nerves to the hypothalamus. Once these messages enter the SCN, a series of physiological reactions takes place.

Q: Is vision necessary for our internal clock to work?

Yes and no. The SCN and the parts of the brain that create images are supplied by different branches of the optic nerves so, technically, we don't need to see for our brains to perceive light or to "tell" time. However, studies show that many blind people have free-running rhythms, rhythms that are not stimulated by how light or dark it is in the environment but by other as yet unknown information. And yet, although their cycles of hormonal production, body temperature, and other physiological activities remain closely related to melatonin levels, they are able to sleep and wake according to a normal eight-hour/sixteen-hour schedule.

Q: What is a zeitgeber?

One theory that explains how the blind organize their internal biological rhythms—and, in fact, how we all manage our circadian cycles despite a fundamental environmental miscue like an electric light—involves the idea of a "zeitgeber." German for "time giver," a zeitgeber is any clue to the time of day presented to your body or your brain. Although sunlight and darkness remain the most obvious and effective zeitgebers, other signs and signals about the time of day flood our environment. The smell of brewing coffee signals morning to many of us, for instance, while the end of the late night news tells us it's time to sleep.

Q: Besides triggering wake-sleep patterns, what other impact does the circadian cycle have on our bodies?

As we've discussed, hormone levels rise and fall in reliable patterns throughout the day (and, indeed, through the week, month, and year). The same is true for body temperature and other physiological activities. It seems that almost every body function has its own window of opportunity in the biological cycle: Body maintenance and repair occur largely at night, short-term memory peaks in the morning, and the senses sharpen in the early evening—all related to the regular rise and fall of hormones. Studies show, for instance, that sensory acuity is highest at 3 a.m., then falls rapidly to a low at 6 a.m., then rises to another peak between 5 and 7 p.m. Cortisol regulates this cycle, and rises and falls in just this rhythm.

Here are some more examples of the ways that circadian rhythms may affect health.

- Asthma attacks are 100 times more common during sleep than during waking hours.
- Heart attacks are twice as likely to occur between 8 and 10 a.m. than between 4 and 6 a.m. or 6 and 8 p.m.
- Mood disorder symptoms peak in cycles as well: Studies show that most suicides occur during the late morning and early afternoon, while the risk of attempted suicide is greatest in the early evening.

Q: Why do most heart attacks occur in the morning?

The exact mechanisms for circadian variations in cardiovascular disease remain unclear, although several factors might apply. Heart rate, platelet stickiness, as well as plasma norepinephrine and epinephrine (the adrenal stress hormones), all peak in the morning hours when the fall of melatonin levels stimulates the release of other hormones, namely serotonin and cortisol.

Q: Could the same principles about circadian rhythm be applied to a life-cycle rhythm?

Yes, and we see it most clearly when tracing the patterns of hormones like HGH, melatonin, DHEA, and the sex hormones. They rise and fall in predictable patterns throughout the life cycle, and each time they do, significant biological events occur, including the development of disease and other side effects of the aging process. Tracking circadian and life-cycle hormonal rhythms is the purview of a new branch of science called chronobiology. At research centers around the world, chronobiologists are studying biologic rhythms as they apply to health and disease, and many scientists believe that chronobiology represents the next wave in medicine.

Q: I've also heard that really low-calorie diets may help increase the life span. Is that true?

Some of the earliest life-span experiments, conducted in the 1930s, showed that rodents fed diets 40 percent lower in calories than littermates allowed to eat freely lived about 50 percent longer and remained more youthful as well, with shiny coats, healthy hearts, and cancer-free cells. A more recent study initiated by the National Institutes of Health in the late 1980s was designed to test the low-calorie theory on longer-lived species such as the squirrel and rhesus monkeys, which normally live between twenty and forty years. Results won't be available for many years, but Dr. Richard Cutler, who is coordinating the study, told a reporter from *American Health* magazine that the dieting animals reach sexual maturity later and appear leaner and healthier than their unrestricted counterparts. One explanation for the effect of calorie-restriction on the aging process involves free radicals: The fewer calories animals consume, the fewer free radicals enter the body.

Q: Besides taking hormones or living on a restricted diet, are there other ways to protect ourselves from the ravages of time?

In a famous study conducted at the University of Southern California School of Public Health during the 1980s, these seven healthy behaviors were found to have a significant influence on mortality rates:

1. consuming only moderate amounts of alcohol
2. eating breakfast on a regular basis
3. maintaining a healthy weight
4. avoiding fat-laden, sugary snacks
5. getting regular exercise
6. enjoying seven to eight hours of sleep a night
7. never smoking cigarettes

On average, men and women who followed these guidelines live about nine years longer than those who had less healthy habits. More importantly, their quality of life at every age was significantly better than that of their peers. They suffered fewer chronic physical ailments and from less depression and other mood disorders.

It's important to keep these statistics in mind as you consider the role HGH or other anti-aging hormones might play in your life: No matter how effective these longevity agents might be, you will still bear the responsibility for your own well-being through healthy eating, exercising, stress reduction, and other habits. In Chapter Nine, we offer tips on making these habits a part of your daily life.

Q: Will it ever be possible to live past 100?
Think of it this way: During the Roman Empire, people usually lived only twenty-two years. In 1850, the average American died at forty-five, and by 1900, by forty-eight. Today, the average age at which Americans die is 75.8, and people over eighty-five constitute the fastest growing segment

of the population. In fact, if current estimates prove accurate, more than two million men and women over the age of 100 will live in the United States by the year 2020.

Add to this the remarkable research taking place in laboratories across the United States and around the world involving HGH, DHEA, and other potential youth-enhancing treatments and the possibility of expanding our life span—and improving the quality of life at every age—does not seem so farfetched. However, we must remember that by attempting to alter the aging process by manipulating hormonal levels, we take two kinds of risks: Risks to our own health by manipulating levels of powerful body chemicals and risks to the entire planet by adding to an already overpopulated and overburdened environment.

Q: Could taking rHGH keep me young forever?

Fortunately, turning back the hands of time simply isn't that easy. I say fortunately because there are many ethical and environmental questions that society must answer before we put an end to the natural process of growing old and passing on. On the other hand, there is every reason to suspect that by finding ways to replace and better balance our hormones as we age, scientists can help us stay healthier and more vital far longer than would otherwise be possible. In Chapter Four, we describe how important strong bones and well-developed muscles are to your general health, and show you the role HGH and other hormones play in their development.

FOUR

Maintaining a Strong Body

Q: What comprises the musculoskeletal system?

As its name suggests, the musculoskeletal system consists of the muscles and bones that form the framework for the body. About 650 muscles, 250 bones, and dozens of joints (junctures of two or more bones) make up the musculoskeletal system. Together, your muscles and bones perform two primary functions: They allow your body to move and they protect your internal organs. Without muscles, we couldn't walk, pick up a pencil, or maneuver a ski slope. Without bones, we couldn't stand upright and our heart, brain, and other internal organs would be unprotected.

In addition, your musculoskeletal system creates the overall shape of your body. Your height and limb length depend on how long your bones grow before and during adolescence and the shape of your body depends on the amount of muscle and lean body mass compared to the amount of fat that sits on the bones.

Q: Why are muscles and bones so important to health, especially as we age?

Most of us take our musculoskeletal system for granted—at least until an injury, disease, or age-related wear and tear prevents it from functioning well. Indeed, the problems that occur when something goes wrong are painful, debilitating, and potentially life-threatening, especially in older people. The prolonged bed rest that results from a musculoskeletal injury or disease often leads to myriad physiological problems, including blood clots, respiratory problems, skin infections, muscle atrophy, and bone loss. In fact, hip fractures caused by weak bones (osteoporosis) represent the twelfth leading cause of death in the United States today: Nearly 20 percent of those afflicted with a broken hip die within three months from complications (such as blood clots and pneumonia) due to prolonged bed confinement.

If your bones and muscles are too weak to hold your body in its proper position, you might experience breathing problems or digestive difficulties as the weight of your upper body collapses upon your chest and abdominal cavity. Perhaps most seriously, if injury or disease prevents you from getting enough exercise, you'll run a higher risk of developing cardiovascular disease, including high blood pressure and coronary artery disease, diabetes, and certain types of cancer.

Q: Is my body shape—the size of my bones and muscle mass—something I inherited? Or can I change my body type?

The answer to both of those questions is yes. As discussed in Chapter Two, body type is a multifactorial characteristic—both hereditary and habits play a role in its development. Several genes, inherited from both your parents, provide a relative blueprint for your musculoskeletal system to follow as it grows. Your height, and to a lesser degree your weight, depends upon the information encoded in your cells' DNA.

At the same time, however, environmental and lifestyle factors also influence the health and longevity of your muscles and bones. Despite inheriting a tendency toward osteoporosis (a weakening of bone tissue), for instance, you can eat a diet rich in certain nutrients and perform specific exercises that will help you keep your bones strong. Proper eating and exercise habits will also help you boost your metabolism (if you're born with a slow one) and thus keep you slimmer than your genes would otherwise proscribe. Exercise will tone and shape even the most limp and undeveloped of muscles—at any age.

Q: How does the health of my musculo-skeletal system change as I age?

As we've discussed, body composition changes as we get older: We naturally gain fat and lose muscle mass. By the time you're sixty-five, your body will contain 18 percent more fat than it did when you were twenty-five, even if you haven't gained a pound. Starting in your twenties, you'll probably also lose about 6.6 pounds of lean muscle mass and a varying amount of bone tissue each decade.

Fortunately, you can improve the health of your musculoskeletal system at any age: In a study performed at Tufts University during the late 1980s, ten frail women and men as old as ninety-six—

some of whom could hardly walk—regularly worked out with free weights. After only two months, the volunteers had boosted their thigh muscle strength so much that they doubled their walking speeds—and two others no longer needed their canes.

Q: What exactly is muscle?
Muscle is body tissue that has the ability to contract, usually in response to a stimulus from the nervous system. The basic unit of all muscle is the myofibril, a threadlike structure composed of complex proteins. These proteins form regularly arranged thin and thick bands of tissue called myofilaments. Each thin myofilament contains several hundred molecules of the protein myosin, while each thick myofilament contains two strands of the protein actin. Myofibrils are made up of alternating rows of thick and thin myofilaments with their ends interwoven. During muscular contractions, these rows of filaments slide along each other by means of cross bridges that act as ratchets.

Three types of muscle tissue exist: smooth, skeletal, and cardiac. *Smooth muscle* is found in the skin, internal organs, reproductive system, major blood vessels, excretory system, while *cardiac muscle* forms most of the heart. *Skeletal muscle*—the muscle we usually think of when discussing issues of fitness and health—is attached to portions of the skeleton by connective tissue attachments called tendons. Skeletal muscle comprises the largest single organ of the body.

Each type of muscle has a different biochemical composition and plays a specific role in keeping the body mobile and the organs functional. Like virtually all other organs and systems, muscles act upon signals sent to them by the neuroendocrine system. Chemical and nerve messengers deliver information to muscles from the brain and other parts of the body, and muscles react by contracting or releasing, or by repairing and building their cells. Cardiac and most smooth muscles receive messages from the autonomic branch of the nervous system, which means we have little conscious control over their actions.

Skeletal muscle, on the other hand, has another name, voluntary muscle, which refers to the fact that we control its movement with conscious thought. We tell our fingers to type on the keyboard, our legs to negotiate a flight of stairs, our abdominal muscles to crunch during a calisthenics class at the gym. Although these activities may seem automatic—we don't have to "think" about every step we take or key we strike—we do, in fact, consciously control the muscles responsible for them through a separate branch of the nervous system.

Q: What is muscle tissue made of?

Like most cells in the body, muscle consists largely of protein. Proteins are large-molecule, nitrogen-containing compounds. Proteins come together to build and maintain muscle through the process of protein synthesis—the creation of protein from raw ingredients called amino acids, of which there are about twenty. Of the twenty amino acids, the body can synthesize twelve; the remaining eight must be provided in the diet and are called *essential amino acids*. Protein deficiencies occur if you do not provide your body with adequate amounts of essential amino acids in your diet.

Once you digest protein, the blood absorbs it through the intestines and sends it to the liver. The liver uses amino acids to synthesize protein for its own use or exports it into the bloodstream. Likewise, the blood will use some amino acids to manufacture its components, then sends others on to the tissues. Each tissue forms a different type of protein from the amino acids: how many amino acids, and in what order they appear, define the type of protein created. Protein in bone is called collagen, while muscle protein is known as myosin.

Q: Where does nitrogen fit in?

Nitrogen is a gaseous element that makes up about 78 percent of the atmosphere and is a part of all proteins and most physical substances. To sustain health, we need to maintain what is known as a *nitrogen balance*: The nitrogen released in the urine, feces, and sweat, together with nitrogen stored in skin and hair, must equal the nitrogen taken in food and drink. Most of the body's nitrogen is blended into protein.

Positive nitrogen balance occurs when the intake of nitrogen required to make tissue is greater than the nitrogen released. Negative nitrogen balance, on the other hand, occurs when the cells lose more nitrogen than they require to maintain their integrity. A negative nitrogen balance, as measured by the amount of nitrogen in urine, indicates that muscles and other tissues are being destroyed. Physicians often look at an individual's nitrogen balance to assess general health and recovery after surgery or other trauma.

Q: What role does HGH play in the process of muscle development and repair?

HGH has an *anabolic* role in muscle metabolism: It helps muscle tissue grow, repair, and replace cells. First, HGH aids in the transport of amino acids from the bloodstream to muscle tissue, and then it helps cells synthesize them into proteins. Once the cells create proteins, HGH helps the muscle cells retain nitrogen, and thus stay strong and whole. Finally, the presence of HGH stimulates the release of IGF-1, a growth factor scientists believe plays a crucial, although as yet little understood, role in muscle metabolism.

Q: I'm not sure I understand what anabolism is. I thought we produced and lost cells all during the life cycle.

You're right. Our cells are constantly dying and being replaced by new cells. Anabolism is one part of the process, catabolism is the other. Anabolism builds and maintains tissues, while the process of catabolism breaks down tissue, produces energy for external and internal physical activity, and helps to maintain body temperature. When anabolism exceeds catabolism, growth or weight gain (usually in the form of lean body mass) occurs. When catabolism exceeds anabolism, such as occurs during periods of starvation or disease, weight loss occurs. When the two metabolic processes are balanced, a state called dynamic equilibrium exists.

Q: What happens to muscle tissue as we age?

As we get older, our bodies become more and more catabolic; that is, they lose cells faster than they replace them. Organs and cells of all types begin to die, releasing nitrogen and becoming smaller and less able to efficiently perform their physiological activities.

Part of the reason muscles—and indeed our bodies in general—become catabolic is that many of us neglect our diets as we get older. We don't eat the protein, fresh fruit and vegetables, and complex carbohydrates we need to replace and renew our tissues. We also tend to suffer more from chronic illness and psychological stress—two conditions that prevent our bodies from either efficiently processing food into amino acids or constructing proteins from amino acids. Finally, and perhaps most importantly, age often brings with it inactivity, and without exercise, muscles and bones begin to atrophy, or slowly weaken and deteriorate.

Q: You keep talking about how exercise builds muscle. How does this happen?

When you add muscle to your body through exercise, you do so by increasing the size of each muscle cell rather than creating new cells. This process is called hypertrophy. Scientists believe that the most effective way to build muscle is by contracting it against added resistance (a process called overload), such as that supplied by weights or even by gravity, which by itself exerts a powerful force against our muscles and bones. Exercise that overloads a muscle causes that muscle to split lengthwise under the strain placed on it. The body then heals the muscle tissue by adding protein to the muscle cells, thereby both strengthening them and adding to their bulk. Over time, especially if you place more and more pressure on the muscles, the muscles will continue to get larger and stronger through this healing process. We know this process as "training."

Q: What role does HGH play in muscle development that results from training?

As we've discussed, HGH aids in the process of protein synthesis, that is, it allows muscle cells to create and assimilate protein. This allows the muscles to repair themselves after exercise and thus develop strength and girth. What's interesting is that the body receives its largest infusions of HGH (apart from the primary nocturnal burst) directly following bouts of exercise. The more you exercise, then, the more HGH flows through your bloodstream, not only helping your muscles to repair and grow, but offering similar benefits to other tissues.

The opposite is true as well, however: If you don't exercise, you'll minimize your supply of HGH and thus put the health of your muscles, bones, and organs in jeopardy. Scientists discovered the link between exercise, HGH, and muscle health by studying humans and animals exposed to environments without gravity, such as those encountered during space flight. Gravity exerts a certain force, a kind of passive resistance, on muscle cells. In effect, simply holding your body in an upright position against the force of gravity represents a kind of exercise. Without gravity to work against, your muscles quickly atrophy. Loss of muscle mass in space is primarily due to decreased synthesis of protein which HGH helps to promote.

Pituitary cells from space flight rats, for instance, secrete only 35 to 50 percent as much growth hormone as rats who stayed on Earth. Laboratory studies suggest that the reduced secretion of growth hormone in gravityless environments is due to decreased sensory input from inactive muscles—in other words, we release HGH when the pituitary receives messages from muscle cells that they are exercising and therefore in need of repair, repair that requires the presence of HGH.

Q: Are there special exercises that help release HGH?

It appears that any type of physical activity that works your muscles (including your heart muscle) will stimulate the pituitary to release HGH. When it comes to building muscle, however, weight-training is the most effective way to increase both HGH levels and muscle hypertrophy. There are three forms of exercise that overload the muscles enough to offer training benefits: isometric, isotonic, and isokinetic. Let's look at each type separately.

- **Isometric exercise**. In isometric exercises, you contract a muscle group without moving the joint to which the muscles are attached. An isometric exercise that works the shoulders, for instance, involves pressing steadily against an immovable wall for ten or twenty seconds. To build chest muscles, you can press your hands together—again, steadily and for several seconds—in front of your chest. By their very nature, isometric exercises use only the human body and fixed objects as "equipment." Instead, your own body weight, and the force exerted by gravity, provides the resistance against which your muscles work. Isometric exercises tend to be very safe, and thus quite useful for older people who suffer from joint diseases such as osteo- or rheumatoid arthritis. However, the results from isometric exercise are more subtle and take longer to derive than other forms of weight training.

- **Isotonic exercises**. Isotonic exercises contract a muscle group through a range of motion, working every angle of the muscles as well as the joints to which they are attached. To perform isotonic exercises, you use either free weights or machines or both. Free weights are barbells (long bars with adjustable weights at each end) and dumbbells (shortened barbells, ordinarily used in pairs, one in each hand). Machines like the Nautilus series provide variable resistance, so that as you move a limb, the resistance stays at or close to maximum.

- **Isokinetic exercise**. Isokinetic exercise uses machines, such as Nautilus equipment, designed to apply maximum stress to the muscles through the whole range of movement. Machines isolate muscle groups very efficiently by maintaining the body in a particular position and guiding the muscles being worked along a specific path.

Q: What kind of exercise routine should I set up if I want to build my muscles?

A typical workout with weights includes a warmup of five to ten minutes followed by an exercise routine that leaves the muscles thoroughly exhausted. You'll want to perform about twelve exercises with weights or weight machines: six for the upper body, six for the lower body. You should program your workouts so that you give each muscle a full day's rest before you exercise it again. If you exercise the same muscles two days in a row, it won't recuperate—HGH won't have time to help the cells synthesize new protein—and thus it will become weaker, not stronger.

Before you attempt to create and follow an exercise routine of any kind, however, you should first talk to your doctor and receive a thorough health examination. That's especially true if you're over forty, unfit, or have a history of cardiovascular disease. Although it may seem as if weight-lifting does not work the heart as hard as aerobic exercise does, it can, in fact, raise your blood pressure to dangerously high levels.

Q: Apart from HGH and IGF-1, are any other hormones important in building muscle?

As we've discussed, the body requires all of its hormones to work together—in harmony and balance—to function with efficiency and strength. Therefore, to some extent, all hormones play a role in the way your body develops and maintains muscle.

That said, perhaps the most important hormone involved in muscle growth and development is testosterone. Testosterone is one of a group of hormones called steroids or corticosteroids. It is an androgen, a hormone that increases the growth of male physical qualities, including muscle development. Both men and women produce androgens, although men naturally secrete much higher levels of these hormones than women. In men, androgens are secreted by the testes and, to a lesser extent, the adrenal glands. In women, most androgens come from the adrenal glands, although the ovaries also secrete small amounts of male hormones.

In men, testosterone is the primary sex hormone, stimulating the growth of the penis and the development of sexuality and sexual functioning. In both men and women, androgens also stimulate other typically "male" characteristics, such as the growth of body hair, the deepening of the voice, and the development of muscle mass. As we age, both men and women produce less and less testosterone, which some scientists think helps explain some of the age-related changes in sexual functioning and muscle mass.

Q: Aren't there other types of steroids?
As mentioned in Chapter One, the adrenal glands produce about thirty other steroid hormones in addition to the sex hormones. Glucocorticoids help regulate the body's use and reserves of sugars and proteins. They also participate in the inflammatory response and the body's reaction to stress. The most important glucocorticoid is cortisol, which plays a major role in protein breakdown and formation, blood sugar control, and reduction of inflammation. Mineralocorticoid is a hormone involved in maintaining salt and water balance in the body. Without this hormone, the tissues and blood would become depleted of salt and water, leading to a potentially fatal drop in blood pressure. Injury and stress trigger the adrenal glands to release mineralocorticoid.

Interestingly enough, DHEA, one of the most recent "anti-aging" remedies, is a steroid hormone that is, in fact, a precursor of other steroid hormones. When the adrenal glands produce DHEA, most of the hormone stabilizes as DHEA sulfate (DHEAS). The liver then converts DHEAS to a number of other hormones, most notably the sex hormones testosterone and estrogen. According to Stephen Cherniske, M.S., author of *The DHEA Breakthrough*, approximately 50 percent of total androgens in adult men are derived from DHEA. In women, the best estimates are that DHEA contributes perhaps 70 percent of estrogens before menopause and close to 100 percent afterwards. But DHEA levels also decrease markedly with age, yet another reason why muscle mass may decline.

Q: Aren't steroids also used as medicine?

Scientists have synthesized hundreds of forms of steroids to treat disease or to supplement hormonal levels in patients who cannot manufacture their own. We're most familiar with corticosteroids as treatments for a wide range of inflammatory conditions. A dab of hydrocortisone cream, for instance, helps to soothe a rash, while an injection of hydrocortisone in the knee joint will relieve the inflammation and swelling of arthritis.

Q: What are anabolic steroids?

As their name implies, anabolic steroids are synthetic preparations designed to increase anabolism, or the building of skeletal muscle tissue. All anabolic steroids are chemical derivatives of testosterone. They work on cells the same way natural testosterone does, by binding with the androgen receptors within the cell. Once in the cell, anabolic steroids act just as testosterone would, that is, by producing anabolic (body-building) and androgenic (characteristically male) effects, such as growth of body hair and deepening of the voice.

Although doctors prescribe anabolic steroids to treat severe wasting diseases and other medical conditions, the biggest market for these synthetic hormones is an illegal one: Today, black market sales of anabolic steroids top $400 million a year. One million Americans, half of them adolescents, use black market steroids to help them build muscle tissue quickly.

Q: Aren't there serious side effects of anabolic steroid use?
Absolutely. Several side effects—some of them fatal—result from prolonged use of anabolic steroids. Because the liver metabolizes anabolic steroids, liver disease, including hepatitis and jaundice, is common with oral steroid use. Steroids also cause negative changes in the blood, increasing its tendency to clot and raising its levels of circulating fats and cholesterol. The presence of anabolic steroids also triggers the adrenal glands to release another type of steroid: the stress hormone cortisol. Excess cortisol often leads to hypertension and neurologic problems, including anxiety and paranoia. In fact, steroid users often experience periods of aggression in a syndrome experts call "bodybuilder's psychosis." Violence against others and self-destructive behavior—including suicide—may result from prolonged steroid use.

Q: I've read that some athletes are now using rHGH instead of steroids. Is that true?
Because rHGH cannot be detected on standard drug screens, more and more athletes are using this

hormone as a way to help them build muscle. Along with rHGH, many also inject a synthetic version of IGF-1, the growth factor triggered by the presence of HGH in the bloodstream.

Unfortunately, the side effects of rHGH and/or IGF-1 therapy may be just as dangerous as those of anabolic steroids, and include joint and muscle pain and swelling, carpal tunnel syndrome, and, if taken in large quantities, deformation of the long bones and the bones of the face. Furthermore, studies show that IGF-1 can trigger the development of breast and colon cancer in some people. These and other side effects caused government officials to amend the Food, Drug, and Cosmetic Act to include this warning: "Whoever knowingly distributes, or possesses with the intent to distribute, human growth hormone for any use in humans other than the treatment of a disease or other recognized medical condition, or such use as has been authorized by the secretary of health and human services under Section 505 and pursuant to the order of a physician, is guilty of an offense punishable by not more than five years in prison, such fines as are authorized by Title 18, United States Code, or both." As you can see, government officials are quite concerned about anyone using rHGH without being closely supervised by a physician. In Chapter Eight, we'll discuss other aspects of rHGH use in further depth.

Q: I'm about forty years old and am starting to worry that my body is losing tone and getting fat. How can I evaluate my body composition?

Physicians and physical fitness trainers use a variety of tools to evaluate body composition. Here are a few you can use.

- **Height and weight tables**. In an attempt to define obesity (and thus calculate clients' risks of developing and then dying from diseases related to obesity), insurance companies developed "Height and Weight Tables," the first one in 1908. The medical community subsequently adopted this table and similar tables for clinical use. In 1959, the Metropolitan Life Insurance Company coined the term "desirable weight for height" and created a table that has since undergone several changes. The problem is that this method doesn't take into account body composition at all, but only measures weight.

- **Body mass index**. Many experts now prefer using the "Body Mass Index" or BMI. This system uses fat, not body weight, as a guide, and attempts to assess the amount of lean body mass (muscle, bone, and organ tissue) compared to the amount of fat in your body. You can calculate your BMI by following this formula: Divide your body weight in kilograms by your height in meters squared. This method is better, but still may classify someone as obese who is very muscular (since muscle weighs more than fat).

- **Skinfold measurements**. A simple and accurate way to assess body fat percentage is through the use of skinfold calipers. To perform this test, your physician or trainer will lift a fold of skin from a number of designated areas on your body. Fat under these areas gives an estimate of the body's total fat tissue. Generally speaking, people whose fat fold measurement exceeds 95 percent are considered obese. The problem with this method is that accuracy depends on the skill of the measurer; an inexperienced technician might be off several percentage points.

- **Underwater weighing**. Perhaps the most accurate way to evaluate both weight and fat/muscle ratio is to be weighed underwater: The more you weigh underwater, the more muscle you have; the lighter you are, the more fat you have. This method is not widely available, but if you're really interested in finding out just how fit you are, search out a fitness spa that offers underwater weighing.

- **Waist-to-hip ratio**. The location of body fat is an important variable in body weight. You can determine the site of fat on your body by measuring your waist's circumference at its smallest point and the circumference of your hips at their widest points, and then calculating a ratio of the two. Depending on your ratio, you may be an "apple," someone who carries excess weight above the waist, or a "pear," someone who carries weight around the hips and buttocks. The higher the waist to hip ratio, the more apple-shaped the figure. Recent research indicates that an abdominal fat distribution pattern is associated with an increased risk for cardiovascular disease. Interestingly enough, older people taking rHGH tend to lose weight around their abdomens, which helps to reduce their chances of developing heart disease.

Q: What about the other half of the musculoskeletal system, the bones? How do I determine their health?

Generally speaking, doctors don't perform routine tests to measure the health of your bones—unless you suffer numerous fractures without good reason (such as a car accident or other trauma) or unless osteoporosis runs in your family. Under those circumstances, doctors can measure the strength and density of bone tissue by performing a bone density scan. In essence, a bone density scan is a sophisticated x-ray that measures a bone's thickness and width. Called photon absorptiometry, this technique emits a fraction of the radiation of x-rays and is sensitive enough to detect a bone loss of just 1 to 3 percent. Another more sophisticated and expensive technique, called quantitative digital radiology, is faster and more precise but is not yet in widespread use.

The doctor will determine the health of your bones by assigning a numerical value to what he or she sees on the x-ray. He or she will then place that number on a graph with the average bone density for someone of your age, height, weight, and ethnic background. If your number falls below that of the average person in your age group and background, you may have osteoporosis, a disease that strikes millions of women (and some men) every year.

Q: How do we normally build and maintain bone tissue?

Although we tend to think of bone as solid material, it is actually living tissue enriched by blood vessels that travel through it. Bone consists of two major layers of cells: the smooth outer surface known as cortical bone and a spongy, meshlike material made up mostly of collagen, the protein-based connective tissue that is also a major component of muscle and skin tissue. When the bone metabolizes the mineral calcium, collagen becomes hard and forms a network of solid tissue or trabeculae. While both trabecular and cortical bone are found throughout the body, their proportions vary from bone to bone. The vertebrae of the spine and the ends of the long bones of the arms and legs, for instance, contain more trabecular bone than other areas of the skeleton, and are thus more prone to breaking and cracking should bones become weak.

Like most other tissue, bone cells are constantly dying and being replaced by new cells. The process of bone regeneration, or remodeling, consists of two distinct stages. During the resorption stage, special cells called osteoclasts dissolve some bone tissue, which creates a small cavity on the bone surface. During the formation stage, cells called osteoblasts take nutrients from the bloodstream to fill the cavities.

The balance between osteoclasts and osteoblasts changes during your life. As an infant and teenager, you have many more osteoblasts than osteoclasts, and hence are able to build bone tissue easily. During your twenties, thirties, and forties, an absolute balance exists between osteoblasts and osteoclasts; you neither gain nor lose bone mass. As you get older, however, the number of osteoblasts decreases while the number of osteoclasts remains about the same. Therefore, your bones lose more cells than they can replace.

Q: I've heard that calcium is important when it comes to the health of my bones. What's the connection?

The primary nutrient contained in bones is calcium, a metallic element responsible for giving bones their strength and density. Calcium is not only important to the health and strength of bones, it is also essential to the proper functioning of many other parts of the body, including muscles, nerves, endocrine glands, and blood cells.

Although calcium is only one factor in bone maintenance, it is an important one. Unlike some substances, such as cholesterol, your body does not produce any of its own calcium. To meet your daily needs, then, you must eat foods rich in calcium or take calcium supplements. Otherwise, you risk losing bone mass, and once lost, bone is almost impossible to regenerate.

Once in the bloodstream, more than 99 percent of the body's calcium eventually winds up stored in bone tissue. Through an elaborate system of hormonal checks and balances, an adequate amount of calcium circulates in the bloodstream at all times; if your body does not receive enough calcium from your diet every day to meet all its needs, however, it will take calcium from your bones to make up the difference. The sex hormones estrogen (in women) and testosterone (in men) appear to help protect bones from being "robbed" of calcium by other parts of the body. That's why women are at such high risk for osteoporosis after menopause.

Q: Are there other minerals important to the health of our bones?

Although often overlooked in favor of calcium, the mineral magnesium is equally vital to bone health. In fact, recent studies show that sufficient magnesium may be even more important than megadoses of calcium alone. A 1990 study published in the *Journal of Reproductive Medicine* showed that supplements containing 500 milligrams of calcium and 600 milligrams of magnesium were sufficient to stop and even reverse bone loss in postmenopausal women. The balance between calcium and magnesium is important to the health of many different tissues of the body in addition to bones, including the heart and blood vessels.

Q: How does HGH affect bone metabolism?

Scientists believe that human growth hormone stimulates the production of osteoblasts, the cells responsible for building bone during the remodeling process. It also increases the amount of a specific protein, known as Gla-protein, that also aids in osteoblast function. It is likely that the age-related loss of HGH may cause or exacerbate the development of osteoporosis.

Q: What exactly is osteoporosis?

The term osteoporosis comes from the Greek *osteo* (bone) and *porus* (pore or passage). With osteoporosis, your bones become too weak to support the body. It occurs because your bones do not receive adequate nourishment to grow and renew themselves nor are they able to efficiently metabolize the nutrients they do receive. In addition, the loss of protective hormones, such as estrogen, DHEA, and testosterone exacerbate the age-related decline in bone health.

Q: Who's most at risk for osteoporosis?
Several risk factors for osteoporosis exist. Let's look at them one by one.

- **Gender**. According to the National Osteoporosis Foundation women are four times more likely than men to develop osteoporosis. Men not only have larger, denser bones to begin with, they also lose testosterone much more slowly and gradually than women do estrogen. Indeed, during the five years following their last periods, women lose bone mass twice as quickly as men of the same age. Today, osteoporosis affects at least half of all American women over the age of sixty. It progresses most after menopause, when bone loss among affected women may occur at a rate of 3 to 5 percent a year. At that rate, they will have lost more than 40 percent of their bone mass by the time they are eighty years old.

- **Age**. The older you are, the thinner your bone will probably be, largely because your body regenerates itself less efficiently as you lose HGH, estrogen, testosterone, and other hormones. In addition, many older people fail to exercise or eat the appropriate amount of vitamins and minerals every day. However, the crippling effects of osteoporosis should not be a normal part of the aging process; osteoporosis is a disease that requires treatment and care.

- **Heredity**. Women and men with a family history of brittle bones are more likely to develop the disease.

- **Ethnicity**. For reasons not yet fully understood, Caucasians of northern European descent, especially those with light eyes and fair skin, have the highest incidence of osteoporosis. It is epidemic among Japanese and other Asian women as well. African-Americans, Hispanics, and women of Mediterranean ancestry appear to be least affected; the incidence of hip fracture is about twice as high among white women as in black women. Jewish women fall in the middle of the range.

- **Physical build**. Because petite women with small bone structures have less bone mass to start with, they are at a higher risk of osteoporosis than large-boned women. A slender build also increases risk; fatty tissue converts adrenal hormones to estrogen, so very thin women tend to have lower levels of protective estrogen than their heavier counterparts. Heavier women are at less risk for another reason: The stress of excess weight on their skeletons stimulates their bones to grow bigger, perhaps by stimulating the release of HGH triggered by the "exercise" required to move the body. (Please note, however, that the health risks of obesity, including diabetes, heart disease, and stroke, far outweigh its minimal protection against osteoporosis.)

- **Sedentary lifestyle**. Exercise is second only to nutrition in maintaining healthy bones throughout the life cycle. Like muscles, bones respond to exercise by growing stronger and larger, thanks to the release of HGH, estrogen, and other hormones stimulated by physical activity. Weight-bearing exercises that cause muscles to work against the force of gravity, such as walking, stairclimbing, running, and tennis, are particularly effective in helping to prevent bone loss.

Q: What happens to bones when osteoporosis strikes?

Osteoporosis attacks mostly the porous trabecular bone, which is especially prevalent in the spine and the long bones of the arms and legs. The holes within the trabecular bone widen, until the bone is eventually unable to support the surrounding cortical shell. A slight fall or blow is all that is needed to fracture a bone beset by osteoporosis.

The bones in the spinal column, called vertebrae, are often the first to become injured. About one out of every four women receives at least one vertebral fracture by the time she reaches the age of sixty; by the age of seventy-five, more than 50 percent succumb to some degree of spinal osteoporosis. Osteoporosis is often the underlying cause of the broken hips suffered by more than two hundred thousand Americans over the age of forty-five every year.

Q: How do estrogen and testosterone affect bone metabolism?

Although no one knows exactly what role sex hormones play in bone metabolism, it appears that the presence of sex hormones stimulates the production of another hormone called calcitonin. Calcitonin helps facilitate the uptake of calcium from the blood into the bone by the osteoblast cells and, at the same time, inhibits the loss of calcium and other minerals from the bone by the osteoclasts. In addition, estrogen helps produce and maintain collagen, an important component of bone, which also decreases in direct relationship to the decrease in estrogen levels after menopause.

Q: Could HGH be used to treat osteoporosis?

We don't know the answer to that question yet. Scientists believe that HGH may help the intestines better absorb calcium and other minerals required to maintain and repair bone tissue. However, studies show that although rHGH increases the rate at which bone cells are replenished, it doesn't help increase bone density, and thus has no effect on strengthening bones weakened by osteoporosis. In fact, one study, published in a 1993 issue of the *Journal of Bone Mineral Research*, showed that low doses of rHGH in elderly patients actually doubled their loss of total bone mass density from 0.7 to 1.8 percent. At this point, then, most experts are wary of prescribing rHGH for osteoporosis.

Q: Can DHEA be used to treat osteoporosis?

As noted, estrogen and progesterone play important roles in maintaining strong healthy bones with advancing age. DHEA is converted into a type of estrogen called estrone in bone cells through the action of an enzyme called aromatase. Estrone increases the activity of osteoblasts, the cells that build bones. DHEA also modulates bone density by affecting osteoclasts as well as osteoblasts. This occurs in the following manner: free radicals are produced by certain immune system chemicals known as cytokines. These free radicals then activate osteoclasts, the cells that absorb bone and make it porous. As people grow older, their levels of cytokines increase, which in turn increases the production of free radicals. Two specific types of cytokines, known as tumor necrosis factor or TNF and interleukin 1 (IL-1) and interleukin 6 (IL-6), are known to increase with age and affect the health of bones.

A study reported in *Nature* magazine shows that TNF inhibits the development of bone collagen and enhances the production of collagenase, the enzyme that breaks down collagen. IL-1 also elevates collagenase, and both cytokines cause resorption of bone. The presence of DHEA, however, appears to inhibit the proliferation of cytokines that normally occurs with age. When scientists inject mice with DHEA, for instance, their immune function—and thus their ability to fight against free radical damage—dramatically improved. Scientists believe that DHEA plays a similar role in this process, but lack definitive proof because they cannot measure cytokine activity in bone tissue with any precision.

Q: Does ERT help postmenopausal women with osteoporosis?
Absolutely. According to the National Osteoporosis Foundation, women who choose to go on estrogen

replacement therapy suffer from 50 to 70 percent of hip fractures compared with women who do not.

Q: In the end, what is the best way to prevent muscle and bone loss as we age?

Without question, replacing hormones like estrogen, testosterone, HGH, and DHEA will go a long way in helping the musculoskeletal system stay stronger throughout the life cycle. We only need look at how effective estrogen replacement therapy is in preventing or at least alleviating the ravages of osteoporosis. However, until scientists learn how to replace these hormones at the right levels and the right times for each individual, most experts suggest improving your diet and getting plenty of exercise, subjects we'll discuss at more length in Chapter Nine.

FIVE

Achieving Cardiovascular Health

**Q: How prevalent is cardiovascular disease
today?**

According to recent statistics, approximately fifty
million Americans suffer from hypertension (high
blood pressure) and about thirty-five million have
some degree of heart disease. Every year, these conditions result in approximately 1.5 million heart attacks and five hundred thousand strokes, of which
about eight hundred thousand are fatal. Most heart
attack and stroke victims are over the age of forty,
making heart disease the nation's number one age-
related condition.

Fortunately, the news isn't all bad: Physicians estimate that in the past twenty years about ten million fewer Americans developed high blood
pressure, thereby preventing some two hundred
thousand strokes and an even greater number of
heart attacks. Advanced medication and surgical
techniques explain much of this decrease, but fundamental changes in the way people live also play
a role. Fewer Americans smoke cigarettes or con-

sume toxic quantities of fat, and more of us realize the benefits of exercise and peace of mind.

Nevertheless, the cardiovascular system remains highly vulnerable to age-related deterioration of tissue. After all, the heart and blood vessels are made of muscle tissue that is just as vulnerable to injury and damage as any other type of muscle. Indeed, simple wear and tear combined with contamination by fats, cholesterol, and free radicals causes the vast majority of cardiovascular disease in people of all ages.

Add to that mix the fact that certain hormones known to protect the cardiovascular system from damage decrease with age—just when we need them most—and you've got a recipe for an intractable enemy like cardiovascular disease. Many scientists believe that if levels of human growth hormone, estrogen, testosterone, DHEA, and melatonin remained at youthful levels, the heart and blood vessels would stay stronger and healthier for much longer.

Q: How does the cardiovascular system work?
The heart and blood vessels form a vast circuit that delivers oxygen and nutrients to body organs and then removes waste products from tissue cells. At the center of the circuit lies the heart. The heart first sends blood to the lungs to pick up oxygen. It then pumps the oxygenated blood to the rest of the body and back through a maze of tubes called the vascular system. Stretched end to end, the blood vessels of your vascular system measure about sixty thousand miles, yet it takes just one minute for blood to complete one full circuit from the heart through the body and back again.

The heart itself is a remarkable muscle. Every year, the heart beats about three million times, which means that by the time you reach the age of seventy, your heart will have beaten more than 2.5 billion times. Here are just a few other facts that may help you appreciate just how hard your cardiovascular system works.

- Each beat of the heart transports oxygen and nutrients to nourish three hundred trillion cells.

- In an average lifetime, the heart pumps one million barrels of blood—and that's only taking into account the work it does when the body is at rest. During exercise or under stress, the heart may pump ten times as much blood as it does at rest.

- The heart beats approximately once a second and sends about 5 quarts of blood coursing through the circulatory system every minute.

- In total area, the capillary walls are equal to about sixty thousand to seventy thousand square feet, or roughly the area of one and a half football fields.

Q: What causes heart disease?
Heart disease takes many forms, from so-called hardening of the arteries to death of the heart muscle (myocardial infarction). Let's take a look at some of the more common forms of heart disease.

- *Arteriosclerotic disease* occurs when fatty deposits block the inside of the coronary arteries, the blood vessels that supply blood to the heart muscle. Angina (chest pain) or heart attacks can occur when the heart's blood supply from the coronary arteries slows or stops.

- *High blood pressure (hypertension)* occurs when the heart's efforts to pump blood meet with higher than normal resistance in the blood vessels outside the heart. A complex interaction among nerve cells, hormones, and chemical cell receptors control blood pressure, and a failure in any of these processes may cause blood pressure to rise. Over time, uncontrolled high blood pressure often damages the heart: If the body's blood vessels are too narrow, the heart must pump harder than normal against the resistance. Eventually the heart enlarges, the muscle thickens, the heart needs more oxygen to function, and it becomes less efficient.

- *Arrhythmia* occurs when the heart's electrical system fails. An arrhythmia can be harmless, such as when the heart beats an extra time, or quite dangerous, such as when actions of the ventricles (the lower pumping chambers) become irregular.

- *Heart muscle disease*, also known as *cardiomyopathy*, robs the heart of its muscle tone and thus leaves it weak and unable to pump blood efficiently.

- *Heart attack* occurs when an area of the heart muscle is so severely deprived of blood that it can longer survive. Heart attacks are also called myocardial infarctions. (*Myo* means muscle, *cardio* means relating to the heart, and *infarct* is a word used to describe the area of dead tissue.) Most heart attacks occur when a coronary artery already narrowed by atherosclerosis is suddenly and completely blocked by a blood clot or muscular spasm. When such a sudden blockage occurs, blood flow to the heart is immediately cut off, causing the death of the heart tissue nourished by the affected artery. Blood clots that form in another part of the cardiovascular system and travel to the heart can also cause the death of heart muscle.

- *Heart failure* occurs when the heart becomes excessively stiff or fatigued from working too hard—either because it must pump too hard against too strong a resistance or because there has been loss of heart muscle strength such as often takes place with cardiomyopathy.

- *Congenital heart defects* are faults in the anatomy of the heart that are present at birth. About twenty-five thousand babies with heart defects are born in this country every year, making congenital heart disease relatively uncommon.

Q: What are the risk factors for heart disease?
High blood pressure and atherosclerotic heart disease represent the vast majority of cardiovascular disease cases in the United States today. Epidemiologists (scientists who study the spread of disease and its symptoms) have identified several risk factors for the development of heart disease.

Risk factors are those conditions and/or habits associated with the increased likelihood of developing a disease. Both risk factors over which you have little control (your age, gender, and genetic background) and risk factors you can change (your diet, exercise habits, and stress levels) influence whether you, as an individual, will develop a certain disease. Let's look at each factor separately.

- **Age**. The risk of cardiovascular disease increases as we get older. More than half of those who suffer heart attacks are sixty-five and older, and about four out of five who die of such attacks are over age sixty-five.

- **Gender**. Men are more likely to develop coronary heart disease, stroke, and other cardiovascular diseases related to atherosclerosis. Scientists believe that estrogens (female hormones) serve to protect the heart and blood vessels far more efficiently than androgens (male hormones). As soon as women lose estrogen's protective effect suddenly and dramatically at menopause, their levels of heart disease quickly match, and even overtake, those of men.

- **Heredity**. Without doubt, some people inherit a tendency toward cardiovascular disease from their parents. In some cases, we understand the genetic pattern responsible, such as when a person inherits a disease like hypercholesterolemia (an inability to metabolize cholesterol). In most cases, however, the inheritance pattern remains unknown.

- **Cigarette smoking**. Cigarette smoking is a major contributor to cardiovascular disease of all types. Overall, experts attribute about 30 to 40 percent of the approximately five hundred thousand deaths from heart disease each year to smoking. Evidence from the Framingham Heart Study shows that the risk of sudden death increases more than tenfold in men and almost fivefold in women who smoke.

- **Obesity**. Any level of overweight appears to increase the risk of heart disease. A study conducted in the 1980s involving more than one hundred thousand women aged thirty to fifty-five showed that the risk for heart disease was more than three times higher among the heaviest group than among the leanest group. And, as mentioned in Chapter Four, the way your weight is distributed may be even more important than how much you weigh. If you have an apple-shaped body, one with more weight gathered in the abdominal area, you run a higher risk of developing heart disease than if your body is pear-shaped, or carrying excess weight around the hips and thighs.

- **High fat, low nutrient diets**. "You are what you eat" is a cliche, but one that has a great deal of resonance when it comes to assessing your risks for cardiovascular disease, particularly atherosclerosis. First, fats and fatlike materials called lipids and sterols are among the substances that attach to injured blood vessels and thus cause the vessels to narrow. Animal fats in particular carry high levels of free radicals, unstable molecules that attack healthy cells. Therefore, the more fat in your diet, the higher your risk for developing heart disease.

Second, certain nutritive substances, including magnesium, chromium, vitamin B6, and niacin help to lower the amount of fat in the blood. Others, like vitamins C, E, and A, and minerals like selenium and zinc, prevent fats and the free radicals in them from being able to damage blood vessels. If you don't consume enough of these substances, your risk for heart disease climbs.

Another important element in any discussion of diet and heart disease is the amount of sugar and simple carbohydrates (like refined white flour) consumed. These substances act on the body in two ways: First, the body stores them more rapidly as fat than complex carbohydrates (like whole wheat flour) or protein. Second, simple carbohydrates tend to raise the level of insulin in the body, which may lead to the development of insulin resistance or diabetes, themselves risk factors for cardiovascular disease.

• **Diabetes and insulin resistance.** Individuals with diabetes mellitus have an increased incidence of coronary heart disease and stroke. The role of insulin, a hormone produced by the pancreas, is to maintain blood sugar at normal levels and to assist glucose in entering each of the body's cells. For some reason, some individuals do not respond as readily to insulin, and more is required to do the job; they have insulin resistance.

Insulin resistance affects the health of the cardiovascular system in several different ways. First, high levels of insulin in the blood may cause the blood vessels to spasm, increasing the risk of heart attack. High levels of insulin cause blood vessels to stiffen, resulting in both high blood pressure (due to the increased resistance) and injury to the blood vessels. Such injury to blood vessels may be an important first step in the development of atherosclerosis, as plaques of fat and cholesterol bind to injured vessels more easily than to healthy ones.

Q: What role does HGH play in protecting the cardiovascular system from disease?

Scientists are now looking at how HGH affects the health of the heart and blood vessels, but as yet are unsure of its exact role. We do know that people born with HGH deficiencies, or who develop an HGH deficiency because of a pituitary tumor, run a greater risk for developing heart disease than people with normal levels of the hormone. Without sufficient HGH, the heart muscle becomes too weak to perform its hard work with efficiency.

As we'll discuss in further depth a little later in the chapter, doctors see a potential role for HGH in the treatment of cardiomyopathy, diseases of the heart muscle itself. HGH may help cardiac muscle grow stronger and more able to perform its vital pumping action. Other research indicates that HGH might help to lower the levels of serum lipids and sterols, thereby reducing the buildup of atherosclerotic plaques. Some people treated with rHGH lost weight in the abdominal region, just where someone at risk for heart disease would want to lose weight. Finally, because HGH helps to strengthen and build all types of muscles, it may help older people stay more active for longer, thus reducing weight-related and inactivity-related cardiovascular problems.

Q: But doesn't having too much HGH raise your risks of developing insulin resistance and/or diabetes?

Yes, and for that reason, many experts caution against using HGH as a supplement without first discussing this very important risk with your physician. Although HGH may well help keep your muscles strong—as you age, taking too much of the hormone may well put you at increased risk for diabetes, a disease related to a host of medical diseases and conditions.

Q: What about DHEA? Should I be taking DHEA to prevent heart disease?

Before we discuss the specific roles DHEA may play in maintaining cardiovascular health, we must stress once again that our understanding of the neuroendocrine system and its effects on health and disease is still incomplete. Although recent studies appear to indicate that replacing essential hormones as we lose them will help us stay healthier longer, we still don't know how much to take and in what combinations, or what the long-range benefits and side effects might be.

That said, recent research indicates that replacing DHEA may be a promising avenue to take for men and women over forty who want to reduce their risk of heart disease, and for several reasons. Recent research indicates that DHEA acts in the following ways.

- **Estrogen releaser**. DHEA is a precursor of estrogen, the female hormone known to protect women against heart disease. After menopause, DHEA helps the adrenal glands turn androgens (male hormones) into estrogen, thereby helping women maintain higher levels of this protective hormone.

- **Insulin inhibitor**. Some research indicates that DHEA may inhibit the ability of insulin to damage blood cells, thus alleviating one of the most harmful risk factors for heart disease.

- **Blood stabilizer**. DHEA may help prevent the blood from forming life-threatening blood clots. In a study performed at the Medical College of Virginia, scientists placed platelets (blood components) and substances known to promote platelet stickiness into a blood vial, then added DHEA. The presence of the hormone significantly reduced the tendency of the platelets to stick together and form clots. Other studies show that DHEA may enhance the ability of the liver to dispose of lipids, lowering the amount of cholesterol and other fatty substances in the blood.

- **Weight loss promoter**. DHEA may promote weight loss in a number of different ways. Some scientists believe that DHEA blocks

the formation of fatty acids that our bodies
would otherwise store as fat, perhaps by inhibiting the action of certain enzymes that
synthesize fat. DHEA also appears to be a
natural appetite suppressant, altering the
brain's hunger signaling system. Recent
studies on animals show that DHEA may
help us lose weight by decreasing our desire
to eat fat.

- **Free radical destroyer**. Both DHEA and another anti-aging hormone, melatonin, are
powerful antioxidants, which means they
fight against the development and action of
free radicals.

**Q: What are free radicals and where do they
come from?**

As mentioned in previous chapters, free radicals
are molecules that contain one or more unpaired
electrons in their orbits. These unstable molecules
destroy healthy cells in an attempt to stabilize
themselves. Our bodies produce some free radicals,
but most come into the body through the air we
breathe or the food we eat. Every cellular process
in the body creates free radicals as a consequence
of using oxygen as an energy source. Hormonal secretion, metabolism, communication between cells,
even protein synthesis all produce free radicals as
a by-product of their processes. The richest source
of free radicals is oxygen: Every time we inhale a
quart of air—the amount we breathe in about thirty
minutes—we expose our cells to one billion free
radicals.

Q: Why are free radicals so dangerous to our health?

The damage done by a free radical goes beyond the first cell it attacks. Not only is that cell injured—its ability to reproduce and metabolize permanently damaged—but it, too, becomes a free radical, ready to ravage another healthy cell. This chain reaction may continue, deforming cell after cell. If the resulting damage occurs to an organ of the endocrine system, certain hormone levels eventually drop, promoting the premature aging of any number of systems and organs.

Q: What is the relationship between free radicals and heart disease?

As discussed, most cases of heart disease result from the progressive narrowing of the coronary arteries that supply heart tissue with the blood and nutrients it needs to survive. Called atherosclerosis, this narrowing occurs when plaques made of fatty substances and blood clots collect on the inner wall of one or both arteries or, indeed, anywhere in the vascular system.

The buildup of atherosclerotic plaque is usually a gradual process. It may take decades before any symptoms of damage appear. The plaque gathers in response to damage to the inner wall of an artery. This damage can come from many sources. Free radicals directly attacking vessel cells, high blood pressure pushing against the arterial walls with greater force than normal, toxic substances such as nicotine and asbestos, and stress hormones all have irritating and damaging effects on blood vessel linings.

Once damaged, the blood vessels attract lipids

and sterols, substances that form the plaques that narrow the vessels and coronary arteries. One of these substances is cholesterol, a fatlike substance known to contribute to the development of plaques and which has an interesting relationship to free radicals.

Q: What exactly is cholesterol?
Cholesterol is a lipid essential for a number of vital processes, including nerve function, cell repair and reproduction, and the formation of various hormones, including estrogen and testosterone and the stress hormone cortisol. Because it is so important to the body, the liver works very hard to create all the cholesterol the body needs to survive. In fact, the liver produces about 3,000 milligrams of new cholesterol in every twenty-four hours, a quantity equivalent to the amount contained in ten eggs.

And there's the rub: The body manufactures all the cholesterol it needs. Any cholesterol you consume is "extra" and can lead to health problems if enough of it circulates in the bloodstream. In addition, cholesterol-rich foods are frequently high in saturated fats and are often fried, leading to the conversion of cholesterol to a dangerous form.

Q: How does cholesterol become dangerous?
Research indicates that it is not the level of cholesterol in your blood, per se, that raises your risk for developing atherosclerosis. Before cholesterol becomes harmful it must be oxidized—or changed into a free radical. Once it is oxidized, it acts as a lure, drawing other cells to it, precipitating a chain reaction culminating in an artery-clogging plaque.

The oxidation process is an intricate one. Cholesterol travels through the bloodstream by combining with lipids and certain proteins. When combined, these substances are called lipoproteins. One type of lipoprotein, called high-density lipoprotein (HDL), is beneficial to the body because it carries cholesterol away from the cells to the liver, where it is processed and eliminated from the body. HDL cholesterol is known as "good cholesterol."

Another type of lipoprotein, however, is considered harmful to the body. Called low-density lipoprotein (LDL), this substance carries about two-thirds of circulating cholesterol to the cells. Current research indicates that LDL cholesterol may become harmful only after it has been oxidized, or combined with oxygen—in essence becoming a free radical itself, a free radical called oxysterol. Oxidized cholesterol enters the bloodstream either from processed foods, ingested animal products (which is why animal fat is so bad for us), environmental pollutants, or from such internal stressors as infections and emotional stress.

Q: What is an antioxidant?
The human body comes equipped with a remarkable set of defenses against free radical damage. They come in the form of antioxidants—enzymes, vitamins, and minerals that mop up free radicals before they can do any harm. Our bodies produce some antioxidants themselves, others we ingest in the foods we eat. Superoxide dismustase, an enzyme found in the liquid part of cells, is an example of an antioxidant the body makes itself. It disarms an antioxidant called dismustase by converting it to hydrogen peroxide, which another antioxidant metabolizes into water.

In addition to these and other chemicals that act directly to protect the body against free radical damage, the body also produces repair enzymes that clean up after free radical warfare. Substances called repair enzymes destroy proteins harmed by free radicals, remove oxidized fatty acids from membranes, and restore free radical damaged DNA.

Despite its complexity, our anti–free radical system is often not strong enough to protect our bodies against the onslaught of unstable molecules that bombard us from the environment. Fortunately, we can reinforce our internal defenses by eating foods rich in antioxidants, taking antioxidant vitamin supplements, and/or replacing hormones known to have antioxidant qualities.

Q: What are the most effective antioxidant vitamins and minerals?

Several studies performed over the last two decades show that antioxidants prevent, or at least slow down, the oxidation of cholesterol. To date, most of the studies involved vitamin and mineral antioxidants, such as vitamin C, vitamin E, and beta-carotene. During the 1990s, experiments at the University of Texas Southwestern Medical Center in Dallas showed that high doses of vitamin E, for instance, rendered cholesterol resistant to oxidation.

Another major study, conducted by researchers at Harvard's Brigham and Women's Hospital in Boston, involved following the diet and health profiles of eighty-seven thousand female nurses over a ten-year period. Investigators found the women whose vitamin E consumption was in the upper 20 percent had a 35 percent lower risk of heart disease, even with all other risk factors—including smoking, family history, and cholesterol levels—considered. Those whose beta-carotene consumption was in the upper 20 percent had a 22 percent lower risk of developing heart disease.

Q: Does HGH have antioxidant properties?

So far, no evidence exists that HGH is a particularly effective antioxidant. On the other hand, a few studies—such as one performed in Italy and published in a 1993 issue of *Acta Endochronologica*—show that rHGH replacement may reduce the oxidized form of cholesterol (LDL cholesterol), probably by helping the liver metabolize LDLs more efficiently. Most current research, however, focuses on the role HGH may play in repairing and then maintaining cells when the heart muscle becomes weakened or damaged for any reason.

Q: My nephew is only 18 and may need a heart transplant because of a virus that destroyed heart muscle when he was younger. What kind of heart disease does he have?

It sounds as if your nephew has cardiomyopathy, a disease in which the heart muscle becomes damaged and fails to pump as often or as hard as it should. Sometimes cardiomyopathy occurs for unknown reasons, but more often hypertension, heart valve disease, artery diseases, or congenital heart defects causes the permanent damage to the heart muscle. There are several different types of cardiomyopathy.

- **Hypertrophic cardiomyopathy**. This disease involves a larger than normal left ventricle. In one form of the disease, the wall between the two ventricles (septum) becomes enlarged and obstructs the blood flow from the left ventricle. Besides obstructing blood flow, the thickened wall sometimes distorts one leaflet of the mitral valve, causing it to become leaky. In over half the cases, the disease is hereditary. Close blood relatives often have an enlarged septum, although they may have no symptoms. The symptoms of hypertrophy include shortness of breath on exertion, dizziness, fainting, and chest pain. Cardiac arrhythmias—irregular heartbeat—may also result, and in some cases, can lead to sudden death.

- **Dilated (congestive) cardiomyopathy**. In this most common form of the disease, the cavity of the heart is enlarged and stretched in contrast to the thickening walls in the hypertrophic form. The heart is weak and does not pump normally, and most patients develop congestive heart failure. This condition often occurs after a heart attack or a buildup of atherosclerotic plaques has already damaged a patient's heart.

- **Restrictive cardiomyopathy**. This is the least common form of the disease. The myocardium of the ventricles become excessively rigid, and the filling of the ventricles with blood between heartbeats is impaired. A person with this type of cardiomyopathy often complains of being tired, may have swelling of the extremities, and may have difficulty breathing on exertion.

Q: Can rHGH used to treat cardiomyopathy?
In 1996, the *New England Journal of Medicine* published an Italian study that described the successful treatment of five patients with severe cardiomyopathy with HGH. The study showed that heart function improved as the cardiac muscle got stronger. Unfortunately, HGH may also affect the sympathetic nervous system (possible by stimulating the release of stress hormones from the adrenal gland). At the dosages provided to patients in this study, HGH triggered arrhythmias (faster than normal and irregular heartbeats). Furthermore, people who suffer from acromegaly (too much HGH) are known to suffer higher risks of developing cardiomyopathy because the heart muscle becomes too large to pump efficiently. Therefore, until scientists determine the right amount of HGH, and the patients for whom the treatment will be most helpful, the role for HGH in the treatment of cardiomyopathy—and indeed for all forms of heart disease—remains under investigation.

Q: Besides taking HGH or other hormonal supplements, what can I do to protect myself from developing heart disease?
Published in a 1992 issue of the *New England Journal of Medicine*, the following statistics about risk factors for heart disease sum up how much potential damage we can avert by complying with sensible lifestyle recommendations.

- **Quit smoking**. Former smokers lower their risk of cardiovascular disease by 50 to 70 percent compared with current smokers.

- **Exercise**. Active people run a 45 percent lower risk than their sedentary peers.

- **Maintain ideal weight**. Men and women who maintain a healthy weight and body composition run a 35 to 55 percent lower risk of developing heart disease than people 20 percent or more above their ideal weight.

- **Maintain normal blood sugar levels**. If you can avoid developing diabetes or insulin resistance, you'll lower your risk dramatically. The risk of cardiovascular disease can be more than six times greater in people with diabetes than those with normal blood sugar metabolism.

- **Replace estrogen (for women only!)**. Women who take estrogen replacement therapy lower their risk by about 44 percent.

Another way you can help reduce your risks of cardiovascular disease and a host of other age-related conditions is to reduce the amount of stress and anxiety in your life. Indeed, your brain, your emotions, and your peripheral nervous system all play significant roles in maintaining your health and vitality throughout the life cycle.

SIX

HGH and the Central Nervous System

Q: I think I understand a little about how the endocrine system works. But what about the nervous system, which works so closely with hormones. How is it organized?

The brain and the nervous system retrieve and transmit messages to and from the rest of your body (using signals from the outside world as well) about movement, coordination, learning, memory, emotion, and thought. The nervous system is divided into two parts: the central nervous system, which consists of the brain and the spinal cord, and the peripheral nervous system, which is the network of nerves throughout the rest of the body.

The spinal cord is a long, almost round structure that reaches from the base of the skull to the upper part of the lower back. The cord carries sense and movement signals to and from the brain and controls many reflexes. The brainstem connects the spinal cord to the brain, and helps control many of your vital functions, such as respiration and circulation. Cranial nerves extend from the brainstem to the muscles of the face, eyes, tongue, ears, and throat, helping to control movement and retrieving messages from these areas to the brain.

Atop the brainstem lie two main portions of the brain, the cerebellum and the cerebrum. The cerebellum helps control coordination, balance, and other unconscious functions. It lies tucked under the cerebrum, which consists of two main portions called cerebral hemispheres responsible for conscious functions such as speech, memory, vision, and some aspects of movement. Scientists have identified specific regions of the brain responsible for certain functions, such as speech and control of muscles in particular parts of the body. For instance, the left hemisphere generally controls the muscles on the right side of the body, while the right hemiphere controls the left side. Other portions of the brain have different responsibilities. The substantia nigra, for instance, is a group of brain cells that produce a chemical called dopamine, which sends messages about movement to and from the brain and the muscles. The thalamus is a collection of nerves that helps to integrate and transmit certain sensations. The brain also contains the hypothalamus and certain endocrine glands, namely the pineal gland and the pituitary gland, which work in close cooperation with the nervous system.

Q: Does the brain control the whole nervous system?

In a way, the brain does lie at the heart of both the central and peripheral nervous systems and acts as the principal "switchboard" for incoming and outgoing messages about emotions, perceptions, and movement. However, current research suggests that the "brain" may be present in many different places in the body.

In particular, scientists have located a kind of "mini-brain" in the sheaths of tissue lining the esophagus, stomach, small intestine, and colon. Neurons, nervous system chemicals, and proteins pack this so-called "enteric (intestinal) nervous system" and, just like the "real" brain, it receives and transmits information to and from the outside world and the internal environment.

The presence of the enteric nervous system helps explain things like getting "butterflies" in your stomach before an important meeting, feeling nauseous after taking an antidepressant or other medication designed to act on the brain, and why ulcers seem to occur more often in people under stress. It also further demonstrates just how complex the nervous system is, and the difficulty scientists face in mapping its activities.

Q: But what exactly is a nerve?

A nerve, or neuron, is the basic unit of the nervous system. A neuron consists of a cell body, one major branching fiber (axon) and numerous smaller branching nerve fibers encased in a sheath. Most nerve sheaths are covered by a white fatty substance called myelin that acts as an electrical insulator for nerve fibers. Made by Schwann cells in the peripheral nerves and by cells called oligodendrocytes in the CNS, myelin keeps nerve impulses from going astray. Some diseases of the nervous system, including multiple sclerosis, involve the loss of myelin, resulting in erratic nerve signals that cause weakness or paralysis, impaired sensation and vision, and loss of coordination.

Q: How do nerve cells communicate with each other?

A neuron receives information in the form of electrical impulses at its dendrite. The impulse first passes through its cell body, then out its axon to other neurons. The axon typically divides into a number of small fibers that end in terminals. The space between the axon terminal of one neuron and the dendrite receptor of another is called a synapse. In order for the electric impulse to pass through the synapse, certain chemicals called neurotransmitters must be present. Neurotransmitters trigger the connection between the axon of one neuron and the dendrite of another.

Stimulation of the nerve sets off an electrical current that releases neurotransmitters stored near the neuron's axon terminals. These chemicals flow across the synapses to stimulate the next neuron in the nerve fiber until the message reaches the brain. Synaptic transmission also includes a process called uptake, in which the nerve endings retrieve some of the neurotransmitter molecules once they've performed their job.

Q: Does HGH have a role in developing or maintaining nerve tissue?

Some animal studies indicate that HGH may play a role in myelination, in which nerves become covered in myelin: Newborn rats rendered growth hormone deficient and those genetically HGH deficient appear to myelinate poorly. When scientists inject growth hormone, these problems are reversed. So

far, studies haven't shown that the same effects occur in humans, however, though research continues.

It seems more likely that HGH's role in the nervous system may be a secondary one. As you may
remember, HGH is responsible for triggering the
release of growth factors known as IGF-1 and IGF-
2. Scientists believe that these growth factors may
affect nerve cells both prenatally and throughout
the lifecycle. Prenatally, they appear to help form
synapses in the brain and at nerve-muscle junc-
tions. Later in life, IGF-1 begins to work primarily
with HGH on maintaining organ and muscle tissue,
while IGF-2 works mainly on nerve cells. However,
it seems that most nerves have receptors for both
types of IGF, which suggests that each plays a crucial role in nervous system maintenence. When levels of HGH diminish, due to aging or disease, the
growth factors precipitated by its presence diminish as well, leaving nerve cells more vulnerable.

Some scientists believe that IGF-1 and IGF-2 may
be part of a group of chemicals called nerve growth
(or neurotropic) factors. One such factor, called
glial cell-derived neurotropic factor or G.D.N.F.
was discussed in four research papers in a 1995 issue of the journal *Nature*. Conducted largely by biotechnology companies, these studies focused on
using G.D.N.F. and similar growth factors for treatment of Parkinson's disease, ALS, and other neu-
rodegenerative diseases.

Q: Are there studies that connect HGH to nerve regeneration?

Interestingly enough, studies performed on a common side effect of diabetes (which too much HGH

can trigger) shed some light on the role of HGH and related growth factors in the nervous system. Called diabetic neuropathy, this painful and debilitating condition affects about 10 to 15 percent of people with diabetes. It involves a gradual deterioration of peripheral nerves, causing a painful burning sensation in the hands and feet, among other symptoms. Eventually, diabetic neuropathy can lead to foot amputations, impotence, incontinence, and gastrointestinal problems.

Until recently, scientists thought the high levels of blood sugar—which they knew often has a damaging effect on neurons as well as blood vessels—caused the nerve cell degeneration. Recent experiments, the results of which have been published in the *Journal of Neuroscience Research*, *Experimental Neurology*, and *Brain Research Reviews*, show that growth factors may be involved as well. Scientists prevented neuropathy in diabetic rats by giving them injections of IGF-2. Although the animals' blood sugar levels remained high, they still got better, which suggests that the hormone and not sugar is more relevant to neuropathy. Clinical trials to test medications that arise from this theory are still a decade or so away, but doctors feel this avenue of research is an extremely exciting one. They also think that nerve growth factors might be involved in other degenerative nerve diseases such as amyotrophic lateral sclerosis, or ALS.

Q: What is ALS?

Also known as Lou Gehrig's disease, after the baseball player who died from it in 1939, ALS involves the progressive degeneration of nerve cells in the brain and spinal cord that control voluntary muscles. The affected nerves shrink and disappear and the muscles they serve to stimulate then waste away. Almost always fatal within ten years, ALS is marked by progressive muscular weakness, atrophy, and abnormal muscle reflexes.

As scientists search for both a cause and a cure, they've made a remarkable discovery. Even as the disease inexorably progresses, the nerve tissue attempts to regenerate and reinnervate itself, actually sprouting collateral peripheral nerve axons—axons that attempt to stimulate muscle fibers whose neural connections were lost. With that in mind, researchers hoped that injecting nerve growth factors (such as IGF-1 or IGF-2) or HGH itself would help the nerve cells complete the process. As described in a 1993 issue of the journal *Muscle and Nerve*, preliminary experiments following this procedure failed. Nevertheless, scientists remain hopeful that some form of effective therapy for this devastating disease will emerge from research on nerve growth factors.

Q: Are brain cells subject to degeneration, too?

Absolutely. In fact, two of the most common diseases of aging—Parkinson's disease and Alzheimer's disease—involve the degeneration of specific cells. In Parkinson's disease, the brain slowly loses cells of the substantia nigra, the tiny orb of tissue responsible for producing an important neurotransmitter called dopamine. Without dopamine, messages primarily about movement do not pass among CNS and peripheral nerve cells with efficiency. Muscles become rigid and movements become slow and limited. Tremors often develop because the loss of dopamine allows another brain chemical called acetylcholine to trigger uncontrolled movement that dopamine would otherwise moderate.

Alzheimer's disease also involves the degeneration of brain cells. In this case, however, the cells involved are largely responsible for thought process and memory. Autopsies of Alzheimer's patients reveal that plaques form around bundles of twisted nerve cells, although how and why this process occurs remains a mystery. Scientists believe, though, that Alzheimer's begins in a portion of the brain called the entorhinal cortex and proceeds to the hippocampus, a waystation important in memory formation. It then gradually spreads to other regions, particularly in the cerebral cortex, the outer portion of the brain involved in language and reason. In the regions attacked by Alzheimer's, the nerve cells degenerate, losing their connections or synapses with other neurons. Some neurons die.

Q: Could HGH or other hormones help?

Without a doubt, but doctors still aren't sure which hormones will work, on which patients, and at what point in the disease process. Nevertheless hormonal treatments of brain disorders remain very fertile ground for research. We've already discussed how HGH and its growth factors might help nerve tissue regenerate in diseases like diabetic neuropathy and ALS, and scientists will continue to examine their potential for treating other degenerative disorders. In the meantime, doctors are already using other hormones to treat CNS disease, as follows.

- **Sex hormones**. In late 1993, scientists first reported a possible link betwen estrogen and Alzheimer's disease. In a study of thousands of women in a southern California retirement community, those who had taken estrogen after menopause had lower rates of Alzheimer's disease than those who had not. In further studies, researchers discovered that certain neurons of the brain have receptors for both estrogen and nerve growth factors (of which IGF-1 and IGF-2 might be a part) and that the presence of estrogen boosts the release of nerve growth factors.

Though far from conclusive, these results prompted two new major studies (still underway) to assess the effect of estrogen replacement therapy in currently healthy older women and in those in the early stages of the disease. Although no comparable studies are planned for men (who do not develop Alzheimer's nearly as often as women), there are hints that older men may benefit from supplements of testosterone, which is converted into estrogen in the brain.

Other research shows a promising role for progesterone, the other major female hormone. Evidence from animal studies shows that progesterone is not only produced by the ovaries, but also locally in the peripheral nervous system, where it promotes the myelin sheath that surrounds nerve fiber. Apparently, progesterone acts in direct coordination with Schwann cells, the cells that produce myelin. In rats with damaged nerve tissue, progesterone levels remained five to ten times higher than blood levels, and increased as Schwann cells and new nerve fibers proliferated and the Schwann cells were actively myelinating the regenerating nerves. Again, exactly how this remarkable research may eventually apply to specific treatments for diseases like multiple sclerosis, ALS, and Alzheimer's disease remains unknown.

• **DHEA.** Since DHEA is actually a precursor for estrogen, progesterone, and testosterone, it should come as no surprise that many scientists believe that replacing this hormone may help older people avoid, or at least delay the onset of, degenerative nerve diseases. Early results on DHEA and memory have been disappointing, though research continues. DHEA is also known as a powerful antioxidant, which means that it might help brain and peripheral nerve cells from being damaged in the first place.

• **Melatonin.** Melatonin, the hormone produced by the pineal gland, also has both antioxidative and potentially regenerative qualities that might help maintain and heal nerve tissue. Recent research shows that Alzheimer's disease patients may have abnormally low melatonin levels, raising the possibility that this disease may be related, at least in part, to the destruction of brain cells by free radicals as we age—destruction that might have been avoided if powerful antioxidant qualities remained available to protect brain cells. The same steady decline of melatonin may also help explain the gradual nature of Parkinson's disease development.

Q: What about pain? How does the brain know what hurts?

Scientists have long been interested in how we perceive pain. No two people experience pain or find relief from it in the same way. Although a fracture of the wrist bone or a cavity in a molar appears to do the same amount of damage in any two individuals, the way each perceives pain may be completely different: One might be in agony, while the other complains of only minor discomfort.

Keep in mind that the word "perception" is an important one in this discussion. That's because the pathway from stimulation (injury) to response ("ouch!") isn't a straight one. When you burn your finger on a stove, for instance, the sensation of pain travels along a series of nerve fibers from your finger to the brain. And there are many ways that the pain message may be altered or even canceled along the route. Have you noticed that when you're in a hurry or preoccupied, stubbing your toe is only a minor interruption, while on other days, perhaps when you're more (or less) relaxed, the same minor injury seems excruciating? Such differences in our own perception of pain point to the complexity of the pain pathways.

Perhaps the most important way that the body protects itself against pain is by producing substances called endorphins. Endorphins are chemicals produced in the brain, spinal cord, and elsewhere in the body in response to the perception of pain or, it appears, emotional stress. Once released, these chemicals attach to certain receptors in the brain and body and act to dull the perception of pain. In fact, opiates such as morphine and heroin have a chemical structure similar to that of endorphins.

Q: I've heard that we release endorphins when we exercise. What's the connection?

The brain produces endorphins under a number of different conditions, including exercise (resulting in the infamous "runner's high"), stress, meditation, and the stimulation of acupuncture points. In addition, certain emotions and emotional responses appear to either trigger or hinder the release of endorphins. Some studies show that depression, which is itself a side effect of chronic pain, decreases endorphin production while laughter increases it.

Q: Is there a relationship between HGH and endorphins?

That's an interesting question that scientists are just now beginning to answer. It appears that when endorphins are released during exercise, their presence may in some way promote the release of HGH. The process probably takes place in the hypothalamus, which interprets high levels of endorphins as a signal to release the healing hormone, HGH.

Q: Does HGH affect emotions?

Apart from its connection to the salving and sometimes euphoric effects of endorphins, HGH probably doesn't directly affect our emotional or psychological lives. However, research suggests that the cause of some, if not most, psychological problems lies in the complex neurotransmission of the brain, involving an imbalance among the brain chemicals and their close endocrine cohorts, hormones like HGH.

Evidence for this theory continues to mount as anti-aging studies continue. After all, the most common psychological condition in the United States today is depression, and depression occurs more in the elderly than in any other age group—including teenagers—and just when their stores of essential hormones dwindle to their lowest point. Some studies show that when we replace these hormones to their youthful levels, depression lifts.

Q: What is depression?
I think the most important thing to understand about depression is that it is more than lingering unhappiness. It is a prolonged illness that leads to a serious disruption in one's lifestyle and activities. Depression involves a pervasive sense of pessimism: Not only are things bad now, it seems they will be bad forever. No external circumstance, such as a visit from a friend or a vacation, will lift the spirits of someone clinically depressed. Other symptoms often accompany this "blue mood," including a disinterest in once favorite activities, work, and sex. Often, thoughts of suicide also arise. Physical changes also occur: Sleep patterns shift, with patients either sleeping much more or much less than before, and appetite either diminishes or grows to an abnormal degree.

Q: How do doctors treat depression?

Doctors now treat most cases of depression with a combination of psychotherapy and drug therapy. Psychotherapy helps patients explore the life situations that may have influenced the development of their disorder (or resulted from it), while drug therapy attempts to address the chemical imbalance that apparently underlies most cases of depression.

Several types of medication help lift depression, each affecting a different set of brain chemicals. One of the most effective types targets a neurotransmitter called serotonin with compounds known as "selective serotonin reuptake inhibitors," or SSRIs. As the name implies, these drugs work by specifically inhibiting nerve cells from reabsorbing (uptaking) the brain chemical called serotonin and thus increasing its availability at brain receptor sites.

Q: What is psychoneuroimmunology?

A relatively new science, psychoneuroimmunology studies the connection between mind and body, specifically the connection between nerve cells and immune system cells. When fighting an infection, for instance, immune cells stimulate the brain to transmit impulses that produce fever. And the reverse is true: when we're under stress, we tend to become ill more often than when our lives run smoothly—and that's because receptors for many of the chemicals released during stress response can be found on immune system cells and organs as well. Psychoneuroimmunology looks at the way the immune system and the neuroendocrine system work together to create and maintain health. In Chapter Seven, we'll discuss the immune system, and HGH's role in maintaining it, in more depth.

SEVEN

The Role of HGH in AIDS and Immunity

Q: What is immunity?
The word immunity comes from the Latin word *immunitas*, which in ancient Rome meant the release of an individual from an obligation to serve the state. Today, we use this word in a similar way. For instance, when known criminals testify in court in exchange for their own freedom, we say they are "immune" from prosecution. To be immune from a disease means you are, in a way, exempt from developing it. Your body has specific mechanisms to protect you from it.

Immunity from a disease develops when a foreign organism or substance, called an antigen, enters the body and is recognized by certain blood cells. This recognition triggers other blood cells into action, which either attack the antigen directly or produce special proteins that neutralize it in other ways. It is an amazing, and still little understood, process that protects us from dangerous foreign invaders and even mutant body cells.

As is true for so many other aspects of health, the endocrine system plays a vital role in the maintenance of immunity. When the body loses certain hormones, including HGH and DHEA, to name just two, the organs and cells of the immune system are unable to efficiently perform their roles.

Q: What organs and cells make up the immune system?

The human body has many ways of protecting itself from harmful substances. First, we have non-specific defenses such as the skin to stop viruses, bacteria, protozoa, fungi, and other disease-causing organisms from entering the body. The body also produces several substances that destroy foreign invaders quickly and efficiently, including the fatty acids of the skin and an enzyme called lysosome found in saliva, tears, and other body secretions.

The organs and cells of the immune system, however, form the main line of defense against pathogens (disease-causing organisms). The human immune system has two components: a fixed compartment of organs and tissues and a circulating compartment consisting of a fluid called lymph and billions of circulating cells and molecules.

The Organs

- thymus (a gland located at the base of the throat behind the breastbone)
- bone marrow (the soft, fatty tissue on the inside of the bones)
- lymph nodes (pockets of tissue located in the mouth, neck, lower arm, armpit, and groin)

- lymphatic vessels (a system of ducts that collect fluid from the body's tissues, deliver it to the filtering lymph node, and transport it to the heart where it rejoins the body's main circulatory system)
- spleen (located in the abdominal cavity near the stomach)

The Lymph

- lymph (a colorless liquid similar to blood but containing no red blood cells or platelets and which carries lymphocytes through the body)
- leukocytes (another term for white blood cells, which help protect the body, produce antibodies, and protect against foreign substances)
- lymphocytes (white blood cells responsible for the immune response; your body contains one trillion lymphocytes, comprising about two pounds of body weight)

Q: How do immune system cells recognize substances as potentially harmful?

The immune system's special characteristic is the ability to recognize other cells in the body. It can tell the difference between harmful microbes, cancer cells, and the body's own healthy cells. Put simply, it's as if every cell and microbe wears a uniform and the patrolling guards—the lymphocytes—can distinguish one uniform from another. Every different kind of cell wears a different uniform, and the immune system responds to each uniform in very specific and particular ways. In addition to harmful microbes and cancer cells, the immune system recognizes transplanted tissues and organs as "nonself," triggering an immune response.

Q: Are there different kinds of lymphocytes?
Yes, there are a variety of immune-system cell types, each with its own set of responsibilities. Natural killer cells, for instance, are large lymphocytes that can destroy diverse microbes on sight, as well as play a role in absorbing cancer cells. Phagocytes and macrophages, which originate in the bone marrow, ingest and digest foreign particles, including some viruses and bacteria.

The division of labor in the immune system is further delineated. There are two kinds of antigen-specific immune responses, closely related and in some cases dependent upon one another. As their name implies, these responses are triggered by specific antigens—cells with specific uniforms recognized by lymphocytes.

Humoral immunity involves the production of protein molecules each time immune cells recognize a new antigen. These molecules are called antibodies. The type of lymphocyte directly responsible for antibody formation is the B cell. B cells are first produced in the bone marrow and have molecules on their surface membranes that correspond to a certain kind of antibody. These B cells circulate in the body until they come into contact with an antigen that corresponds to its antibody structure. The antigen then locks onto the B cell, which provokes an immune response. The B cell rapidly and repeatedly divides, creating hundreds of new cells that release antibodies in massive amounts.

Antibodies are antigen-specific, able to react only with the antigen that produces them. B cells also produce memory cells, cells stored in the lymph that have the ability to recognize the specific antigen that provoked their production. Should this antigen reenter the body, memory cells quickly divide and release appropriate antibodies immediately.

Q: What are T cells?
T cells perform another kind of immune system response called cell-mediated immunity. The bone marrow also produces T cells, but then T cells make another stop in the thymus. The thymus takes immature lymphocytes and processes them into T cells (thymus-derived cells). T cells also have molecular codes on their surfaces that correspond to specific antigens. When T cells recognize and bond to antigens, cell-mediated immunity occurs. Some T cells—appropriately called TK for T-killer cells—release a lethal poison as soon as they come into contact with certain antigens.

The rest of the T cells consist of T-helper cells and T-suppresser cells. These cells work together with other immune system cells to create the immune response. T-helper cells, as their name implies, call macrophages and TK cells into action and stimulate B cells into antibody production. T-helper cells are the "on switch" for the fight against many microbial infections.

T-suppresser cells represent the "off switch." When a crisis ends and the invading microbes have been destroyed, T-suppresser cells signal to other lymphocytes that the danger has passed. Macrophages move away, TK cells stop reproducing, and B cells halt their production of antibodies.

T cells communicate these messages to other lymphocytes by releasing enzymes, peptides, and proteins, and by other means as yet unknown. What we do know, however, is that when either the cell-mediated or humoral branch of the immune system malfunctions, we become vastly more vulnerable to disease.

Q: Does HGH affect the immune system?
Studies show that the presence of HGH may help the immune system perform its vital work. Like many other hormones, including melatonin and DHEA, HGH appears to directly enhance the activity of the NK cell, the large molecules that directly attack cells the immune system recognizes as "nonself." The presence of HGH also stimulates the thyroid gland to release its hormone, thyroxine. In turn, thyroxine triggers the thymus gland into action.

Q: What is the relationship between the thymus and HGH?
The thymus gland is a primary organ of the lymphatic system that produces enzymes that govern the production of T cells. As we age, the thymus begins to shrink: In fact, the organ is largest relative to the rest of the body when we are about two years of age, and attains its greatest absolute size at puberty. After puberty, however, the organ shrivels as fatty tissue replaces the receding thymic tissue. By the time we reach the age of sixty or so, the thymus usually appears as tiny islands of thymic tissue covered with fat. In addition to stimulating

the remaining cells of the thymus to produce T cells, the presence of HGH also releases the growth factor IGF-1, which acts directly on the thymus to produce both T cells and B cells.

Q: What happens when the immune system malfunctions?

There are two general categories of immune system diseases: autoimmune and allergic conditions and immunodeficient diseases.

Although quite different in their causes and their effects, physicians consider allergies and autoimmune diseases to represent overreactions of the immune system. Most of the time, the immune system works efficiently to protect the body from harmful invaders. With allergies, however, the immune system misreads the signals and responds to substances that are actually benign. Such reactions involve an interaction between a specific foreign substance (pollen, for instance) called an allergen, and a specific antibody. When an antibody attacks an antigen, it releases a substance called histamine, a body chemical that can act as an irritating stimulant, and other chemicals into the tissue. These chemicals act to produce what we think of as the allergic response: runny nose, sneezing, itchy skin, rashes and/or shortness of breath, among others.

Autoimmune diseases, on the other hand, involve a mistake of a different kind on the part of the immune system. In this case, immune system cells target certain cells in the body for destruction—for reasons completely unknown. There are two general categories of autoimmune diseases: collagen diseases that involve the connective tissues, such as rheumatoid arthritis, which affects primarily joints, and systemic lupus erythematosus, which may attack organs and nerve tissue; and vascular diseases, such as hemolytic anemia, in which antibodies destroy red blood cells.

Q: What is AIDS?

Acquired Immune Deficiency Syndrome is caused by a virus known as HIV (Human Immunodeficiency Virus). The virus attacks a very specific subset of lymphocytes, namely the T-helper cells. Without T-helper cells to stimulate the rest of the immune system into action, the body is vulnerable to a host of potentially life-threatening and painful infections—infections the body normally would be able to fight pretty easily if the immune system was intact. Almost fifteen years after scientists first isolated HIV and identified the disease process, the search continues for ways to solve what is now a worldwide health crisis. To date, epidemiologists (scientists who study the progress of disease) estimate that HIV infects more than one million Americans. More than 250,000 men and women have died of the disease since 1981, some four million people around the world have confirmed cases of full-blown AIDS, and another seventeen million to nineteen million may be infected with the virus.

Q: What happens to T cells when HIV invades?

HIV is a retrovirus, a virus that contains within its core a different kind of genetic material called ribonucleic acid, or RNA. When the retrovirus attaches to the T cell, it uses a special enzyme, called reverse transcriptase, to convert RNA into DNA.

DNA, as you may remember, is the material that carries all the genetic information needed for a cell to function and reproduce. Once this occurs, the next generation of that T cell will also include the genetic material for the virus, and so will all future generations. Once a retrovirus infects a person, he or she is apparently infected for life, even if no symptoms present themselves. That's why the incubation period for HIV may last as long as several years. However, once activated by a concurrent infection or unknown trigger, the virus may already have infected vast numbers of once healthy T cells.

A virus, especially one as "tricky" as HIV, is perhaps the toughest microbe to fight once infection occurs. It reproduces within a *healthy* cell, using that cell as protection against the body's own immune system and any medicines that might kill it. In fact, to kill a virus once it has infected a healthy cell usually involves killing the healthy cell as well. After years of infection, the T cells of AIDS patients become severely depleted and the viral load (the amount of virus in the system) increases.

The existence of an immunodeficient condition like AIDS leaves the body opens to infections that it could otherwise fight off. Some can be lethal, such as certain cancers and pneumonias. Others are simply painful and annoying, such as rashes, persistent fever, and swollen lymph glands. Infections that attack the body when an immune system is not functioning properly are called opportunistic infections. They take the opportunity to attack when the body is unable to fight back.

Q: What is AIDS wasting?
AIDS wasting syndrome is defined as a loss of at least 10 percent of body weight. No one knows for

sure what causes this wasting away, which also occurs in other serious conditions, including some cancers. In AIDS patients, wasting may be lethal.

Until recently, most AIDS patients with wasting took appetite stimulants. Unfortunately, although appetite improved and patients gained weight, they gained only body fat. Their muscles and organs continued to lose nitrogen and waste away. Two FDA-approved drugs, both appetite stimulants, worked to increase a patient's body fat rather than lean body mass. Scientists developed Serostim, a type of rHGH, specifically to address AIDS wasting.

Q: How does rHGH help AIDS patients suffering from wasting syndrome?

Scientists believe that rHGH helps the body to synthesize protein more efficiently, just as it does for children and adults suffering with HGH deficiency, thereby repairing and maintaining lean muscle mass and cell membranes. Improved protein synthesis allows them to gain weight in the form of muscle and organ tissue rather than fat.

A 1994 study performed at San Francisco General Hospital involved 178 volunteers: men and women who had lost 10 percent of their pre-illness weight, or to weigh less than 90 percent of their ideal body weight. In the trial, ninety volunteers received growth hormone (an average dose of 6 milligrams per day administered subcutaneously), and eighty-eight received placebo for three months.

During the trial, patients in the placebo group initially gained an average of one pound, but proceeded to lose most of this gain during the following three months. Those in the HGH group, on the other hand, gained an average of more than 3½ pounds, and have sustained or increased this gain after the trial. Several patients eventually gained more than 20 pounds while on extended treatment. More importantly, their weight gain consisted not of fat, but of lean muscle mass.

In addition, preliminary studies indicate that rHGH helps reduce the likelihood that a cancer of the blood-making tissues, called myeloblastic anemia, will develop from treatment with AZT, an anti-AIDS drug. Again, doctors are unsure of why HGH seems to help, but continue to perform studies.

Q: Do people with AIDS run any special risks if they use rHGH?

According to lab studies performed by Serono Laboratories, the pharmaceutical company that makes rHGH for AIDS wasting, rHGH has been shown to stimulate replication of HIV. However, no such replication occurred if antiviral drugs were added to the culture medium. Therefore, an AIDS patient on rHGH should be taking an antiviral medication such as zidovudine or didanosine—a matter he or she should discuss thoroughly with a qualified physician.

Other than this potential complication, AIDS patients taking rHGH experienced the common side effects of rHGH therapy, including swelling of the hands and feet, musculoskeletal discomfort, and carpal tunnel syndrome. However, these relatively minor problems—many of which resolve spontaneously after a few weeks—may well be worth the benefits the drug provides patients suffering from the physical devastation of wasting.

Q: Does DHEA have a role in maintaining the immune system?

DHEA, the hormone secreted by the adrenal glands and a precursor of estrogen and testosterone, appears to restore immune balance and stimulate production of several different immune system cells, including the following.

- **Monocyte production**. Recent research shows that DHEA administration increases production of monocytes, the cells that directly attack tumor and other cells identified as "nonself."

- **B cell activity**. In the presence of sufficient DHEA, B cells react more swiftly and powerfully to defend the body against a large number of disease-causing organisms.

- **T cell mobilization**. Scientists have found DHEA-binding sites on T cells, meaning that the presence of DHEA helps T cells effectively function in coordination with the rest of the immune system. A landmark study published in the *Annals of the New York Academy of Sciences* found that supplementing elderly individuals with DHEA increased T-cell production of two immune substances called interleukin-2 and gamma interferon.

- **Thymus gland protection**. As discussed, the thymus gland represents an important source of T cells. DHEA helps to protect thymus gland cells by moderating the effects of stress hormones (epinephrine and norepinephrine), which accelerate the demise of the thymus. DHEA also stops stress hormones from interfering with the immune system's ability to turn on specific genes to mount an effective immune response.

Q: Does melatonin have a role in boosting the immune system?

Scientists are still investigating melatonin's role in the maintenance and functioning of the immune system. They're currently following up on early research that identified melatonin-binding sites—receptors on the surfaces of cells that attract and attach melatonin—in the thymus and the spleen, two important immune system organs. We've also seen that pinealectomy—the removal of the pineal gland and the resulting loss of melatonin—induces a state of immunodepression in rodents, but when melatonin is replaced, immune system function improves dramatically.

Studies performed at the Department of Molecular Pharmacology and Biologic Chemistry at Northwestern Medical School found that melatonin activated monocytes and enhanced their ability to destroy nonself cells. The same study showed that melatonin influenced the release of IL-1 (interleukin-1), proteins that control aspects of blood cell production and the immune response. One animal study of particular interest showed that melatonin could restore the function of T-helper cells in mice whose immune systems had been compromised. The results of this study may someday lead to improved treatment for humans with AIDS.

Q: Apart from viruses like AIDS, what can cause an immune system to break down?

The immune system may malfunction or become depleted for a number of reasons. Certain conditions are present at birth and result from a genetic defect. Nutritional deficiencies trigger other immune disorders, and infections by fungi, bacteria, and other viruses may also attack immune system cells or organs. Chemotherapy and irradiation—common treatments for cancer and other illnesses—often suppress immune function. Severe burns and other traumas to the body, including surgery, are also known to disrupt our ability to protect ourselves from disease.

One of the primary depressors of the immune system is stress, or more precisely, the hormones the body produces in reaction to stress. Produced by the adrenal glands, the stress hormones are involved in what is called the "fight-or-flight" response. When your body is threatened in any way, it mobilizes immediately, preparing you either to battle the impending danger or to flee from it. To do so, the adrenal glands release stress hormones into the bloodstream that, in turn, trigger organs and tissues of the body to action.

Two of the most powerful stress hormones, norepinephrine and epinephrine, work to provide a fast and usually short-lived response to an immediate threat by stimulating the nervous system to raise the blood pressure, increase the heart, respiration, and metabolic rates, and prepare the muscles for action. Later, the adrenals release other hormones to continue fighting after the effects of the fight-or-flight response are over. These hormones, called corticoids, include cortisol and aldosterone.

The reactions induced by the fight-or-flight response during and after we sense danger are essential for survival. However, problems occur when these powerful hormones and the reactions they stimulate continue over a long period of time due to perceived emotional or psychological pressure. As discussed in Chapter Five, cortisol elevates blood sugar, increases platelet stickiness, and elevates serum cholesterol, all of which cause or exacerbate age-related heart disease and diabetes.

Chronic stress also depletes the immune system. Studies show that after periods of extreme stress, T-helper cells do not work efficiently and thus cannot protect us from infection. Stress hormones also trigger the thymus gland to shrink, thereby decreasing one of the most important storehouses of T cells.

Q: Are there substances in the environment that can affect the immune system?

As is true for virtually all other body cells, lymphocytes and leukocytes are targeted for destruction by free radicals. Free radicals come from a variety of environmental sources, including polluted air, electromagnetic fields, cigarette smoke, and ultraviolet rays from the sun. In fact, researchers at the University of Michigan at Ann Arbor performed a study on the immune system. They found that skin that had been sunburned showed less reaction to allergy-causing agents than healthy skin—a sign that the immune system (which also produces allergic responses) was not functioning properly. The results of this study indicate that overexposure to ultraviolet rays can be harmful in two ways: first, by stimulating the creation of cancer cells by damaging DNA and second, by inhibiting the immune system from recognizing and

eliminating skin cancer cells once they are produced.

**Q: What else can we do to boost the immune
system?**
As is true for so much about health and healing,
any habit that is good for one part of your body
will help your immune system stay healthy as well.

- **Eat a balanced diet**. Like all other body tissue, the immune system needs protein, vitamins, and minerals in order to maintain its
  health. Severely malnourished people are
  particularly vulnerable to immune system
  dysfunction.

- **Exercise in moderation**. Without question,
  regular and vigorous exercise is one of our
  most important health- and longevity-
  enhancing activities. Some studies show that
  exercise induces a short-term rise in the
  number of immune system cells. However,
  other research indicates that rigorous, continuous exercise may suppress immune
  function over the long term. Chances are,
  though, that even an ambitious exercise regimen will do far more good than harm. If
  you harbor the human immunodeficiency
  virus or another immune-depleting disease,
  check with your physician before undertak-
  ing an exercise program.

- **Get seven to eight hours of sleep a night.**
  Little conclusive research has been done on
  the effects of sleep on the immune system.
  However, we do know that the release of

HGH that comes with sleep helps the body repair itself, and that includes the cells and tissues of the immune system.

In Chapter Eight, you'll learn more about the many hormones now available as supplements that may help you keep your body up and running—efficiently and vigorously—even as you age.

EIGHT

Using HGH and Other Supplements

Q: I'm interested in taking human growth hormone and maybe other anti-aging supplements. What's my first step?

First, consider your age and general health. If you're under fifty and in fairly good health, it might be too early to start taking supplemental or replacement hormones since your body probably produces enough on its own. The best thing you can do now for your present and future health and longevity is to exercise on a regular basis, eat a balanced diet, sleep well, and manage your stress. These habits will not only boost your general health, they will also help your body produce optimum amounts of HGH even as you age. In fact, exercise appears to be one of the most effective hormone-releasing stimulants available: Regular, vigorous exercise triggers the release not only of HGH, but also of DHEA, estrogen, testosterone, and melatonin. If you're older than fifty, now might be the time to consider replacing some of the hormones your body is now beginning to produce in lesser quantities—especially if you lack energy or experience any other side effects of aging, such as loss of libido. In this chapter, we discuss how to evaluate your need for these hormones and how to take them.

If you're seriously ill at any age, especially with a disease such as HIV that involves wasting or a disease that may lead to heart failure, you might want to consider taking HGH and/or another hormone supplement. That's a matter for you to discuss with your physician.

Q: Are there people who should not take hormone supplements?
There are several medical conditions that might be complicated by taking hormone supplements, and thus if you currently take any medication or are undergoing other medical treatment, talk to your doctor before you take a hormone supplement. That said, there are several other categories of people who—in general—should not take hormone supplements, including those who are trying to become pregnant, pregnant or nursing, suffering from kidney, endocrine, or autoimmune diseases, or under fifty and healthy.

Q: I'm 65 years old and suffer from type II diabetes and high blood pressure. I take medicine for both conditions. Should I take rHGH or other hormones or would they interfere with the drugs I'm taking?
You're right to be concerned. Both insulin for diabetes and drugs used to treat high blood pressure directly impact the endocrine system and may thus interact with rHGH and other hormone supple-

ments. Many other medications for other ailments do so as well. That's why it's important for you to work closely with your doctor should you decide to take supplements. On the other hand, so far, studies on HGH, estrogen, testosterone, melatonin, and DHEA show no adverse side effects or drug interactions when taken exactly as directed by a physician and closely monitored. Again, please discuss the matter with your physician *before* you take any supplements and DO NOT stop taking your medication for any reason without your doctor's explicit recommendation.

Q: Are there any tests that will help me decide if I need to start taking HGH and other hormones?

At this point, hormone replacement is a highly inexact science, even when it comes to a tried-and-true hormone like estrogen for women entering menopause. Although we can measure blood levels of most hormones with a simple series of blood tests, we're not sure what the resulting measurements may mean to our long-term health, and we certainly don't yet have a precise and foolproof way to replace them all so that the endocrine system functions in its former youthful balance.

Another thing we don't yet know is the effect of replacing just one hormone among the many others that decline with age. Could replacing only HGH while allowing DHEA or testosterone to decline somehow throw the body out of whack? How do we know which hormones might work best for us? Should people take a specific combination of replacement hormones tailor-made for their own particular body chemistry? These are just a few of the questions anti-aging scientists are currently exploring.

In fact, many experts are convinced that someday soon scientists will develop a "superhormone cocktail," containing appropriate doses of several different hormones to replace those we lose through the aging process. Until then, it's best to work with your doctor to decide which hormones you would most benefit from taking. Your first step should be to undergo a complete physical with laboratory tests. Among the tests your doctor should give you are the following.

- **CBC**. A complete blood count will check for anemia and signs of infection.

- **Cholesterol/Lipid profile**. Your doctor should do a full lipid profile in order to measure your total cholesterol level, your ratio of HDL (high density lipoprotein, the "good" cholesterol) to LDL (low density lipoprotein, the "bad" cholesterol), as well as your triglyceride levels.

- **Liver function test**. Your liver does a great deal of work to keep your body up and running, especially when it comes to metabolizing medications, vitamins, and hormone supplements taken by mouth. (Most substances injected into the bloodstream or that reach the bloodstream through a transdermal patch bypass the liver, including HGH.)

- **Thyroid function test**. Your thyroid helps control metabolism; when it malfunctions, it can cause a host of symptoms and side effects, including unwanted weight gain and depression. Hyperthyroidism (overactive thyroid function) and hypothyroidism (underactive thyroid function) are often overlooked, especially in the older population.

- **Pap smears and breast exams**. These tests are especially important for women taking estrogen, as estrogen can stimulate the development of reproductive cancers in some women.

- **Digital exams for prostate and colon cancer**. Men should also have a PSA test (prostate-specific antigen test) which can help detect any latent prostate tumors. These tests are especially important for men taking testosterone, as testosterone can stimulate the development of prostate cancer in some men.

As for specific hormones, your doctor can order blood or saliva tests to measure the amount of hormones in your body. Estrogen, testosterone, and DHEA levels can be evaluated pretty simply in this way. HGH and melatonin, however, are more difficult to measure, since the body produces them in spurts and levels vary greatly throughout the day. If you're over fifty, however, you can assume that your body no longer produces youthful levels of these hormones and thus would benefit from supplementation.

Q: My doctor is very conservative and I'm not sure he'll support my taking anti-aging hormones. Should I tell him about it?

I'm not surprised that you feel you might meet some resistance to taking HGH or other hormones.

When it comes to health issues, most doctors trained in western medicine tend to look upon prevention as secondary to treatment. In other words, your doctor might feel more comfortable suggesting you take calcium supplements to treat your osteoporosis or prescribing anti-hypertensive drugs to treat your high blood pressure rather than offering hormonal supplements or other natural alternatives.

Ask your doctor if he would consider prescribing HGH or supporting your interest in taking other hormones. Bring this book and other information to your next appointment so the doctor can start to evaluate the research into the effects and side effects of these substances for himself. If your doctor simply refuses to consider the use of hormone supplements of any kind, you may want to change physicians. Such a decision should not be made lightly, of course, especially if you've been under your current physician's care for some time. On the other hand, only by working within an atmosphere of trust and mutual respect will you receive the health care you deserve.

Please be advised, however, that you *must* let your doctor know if you're taking hormone supplements, especially if you're already taking medication for an existing condition.

Q: How concerned should I be about side effects of these hormones?
You should be quite concerned, especially when it comes to a hormone like HGH, which is difficult to measure or prescribe in accurate doses. Before you take any type of hormonal supplement, you should weigh three major issues.

- **Risks vs benefits**. Find out as much as you can about the potential benefits to your body and weigh these benefits against the suspected risks. The potential benefits of HGH replacement are impressive: Studies show that rHGH can help maintain bone density and muscle mass, alleviate some types of heart disease, reduce body fat, aid in healing surgical and other wounds, and boost the immune system. But HGH has its risks as well, including the potential for triggering the development of carpal tunnel syndrome, edema, abnormal bone growth and—in the long term at least—perhaps even some cancers. Furthermore, rHGH may exacerbate diabetes and other endocrine disorders.

  Someone who has diabetes, for instance, may decide that the potential side effects of rHGH, which include potentially exacerbating insulin resistance, outweighs its benefits. A healthy woman in her fifties at risk of osteoporosis and with a tendency to gain weight, on the other hand, may conclude that the potential of rHGH to maintain bone density and to reduce fat outweighs its side effects.

- **Cost**. When it comes to your health, you may feel that money is no object, and rightly so. However, rHGH remains prohibitively expensive for the vast majority of us, costing at least twelve thousand dollars a year—and, unless you suffer from a diagnosed growth hormone deficiency or wasting syndrome, your insurance won't cover it. Other hormones, such as melatonin and DHEA, tend to be quite inexpensive, usually less than fifty dollars a year.

• **Availability**. HGH is not yet available over the counter, and probably will remain a "by prescription only" product in the foreseeable future. Furthermore, your own personal physician may not be willing to prescribe the drug for you for anti-aging purposes. However, in our resource section at the back of this book, we list two organizations that can help you find a physician who will work with you to replace HGH.

Q: How do I take rHGH?
After a doctor examines you and performs standard laboratory tests (such as those listed above), he or she will then teach you how to give yourself injections of the drug and provide you with a month's supply of premeasured syringes. Although regimens vary from doctor to doctor and patient to patient, a standard regimen consists of four to eight IUs (International Units) of rHGH taken in twelve doses over six days. Usually, you'll give yourself two shots a day, one in the morning and one at night.

Q: What can I expect if I take rHGH injections?
If your experience is a typical one, you'll probably feel more energetic, enjoy deeper sleep, and notice your moods are brighter and you're better able to cope with stress within just a few weeks of starting treatment. During the second month you use rHGH, you may notice your muscles firming up, your body fat disappearing, and your overall en-

ergy and vitality increasing. Some studies indicate that you may even experience a decrease in total cholesterol within a few months. By the time you've used rHGH for three months, you should notice a definite improvement in lean body mass, muscle shape and tone, continued loss of fatty tissue, and further improvement in skin thickness and elasticity.

Q: How long do I have to take rHGH?
If you want to use rHGH for anti-aging purposes, research indicates that you must use it on a continuous basis for the rest of your life in order to maintain its benefits. Once you stop injecting yourself with the hormone, you'll start to see fatty tissue return and lean muscle mass decrease within just a few months. However, if you are ill or injured and your doctor prescribes it as a way to bolster your body during this stressful time, a shorter course of treatment may be possible.

Q: What side effects occur with rHGH use?
As discussed in Chapter One, most people taking rHGH experience relatively mild and often temporary side effects, as least in the short term. However, if you decide to take rHGH, you and your doctor should keep a careful watch for signs of the following.

- **Allergic reactions**. Approximately one-third of all patients taking synthetic HGH develop antibodies to it. As you may remember from Chapter Seven, antibodies are immune system cells created to destroy substances the body considers "enemies." In addition to causing rashes near the site of the injections, the creation of antibodies to rHGH may deplete the immune system.

- **Carpal tunnel syndrome**. This serious joint disease occurs when pressure on the carpal nerve, which runs through the wrist to the elbow, causes weakness, pain when the thumb bends toward the palm, and burning.

- **Abnormal bone growth**. Also known as acromegaly, this disorder occurs when HGH stimulates the bones to grow after adolescence. Because the ends of the bones have been fused, and thus cannot grow lengthwise to add height, instead the bones grow in width, creating facial and other deformities.

- **Disturbed insulin metabolism**. HGH causes the body to use fat instead of glucose as energy, which produces abnormally high blood sugar levels. This condition can lead to hyperglycemia or diabetes.

Q: I've heard about tablets of human growth hormone sold in health food stores or over the Internet. Are they useful?

Not at all. To date, the body responds only to injections of HGH. Any over-the-counter pills or liquids claiming to be HGH may either be useless or even downright dangerous. Do not take HGH in any form unless you're working directly with a physician who prescribes it to you.

Q: Are there other supplements that will stimulate the release of HGH?

As discussed in Chapter Two, amino acids—substances that make up proteins—are required for the body to produce HGH. Of the approximately eighty amino acids found in nature, several appear to directly stimulate the pituitary gland to produce natural growth hormone. The three most active amino acids are l-arginine, l-lysine, and l-lysine's close cousin, ornithine. Studies show that when taken in combination, these amino acids may stimulate HGH production.

Unfortunately, it appears that the amount of active amino acid available in most oral supplements now available in over-the-counter preparations are too small to have any effect on the pituitary gland's release of HGH. In addition, it would be difficult—and perhaps even dangerous—to attempt to self-prescribe amino acids, since overdoses of these substances can cause serious side effects. You must carefully combine and time the administration of amino acids in order to trigger the pituitary to release HGH. The good news is that it is relatively easy to provide the body with all the amino acids it needs by eating enough protein every day. In Chapter Nine, we discuss how much protein you need and what foods to choose.

In addition, a number of scientists in several laboratories across the country are now working to develop one or more drugs that will stimulate HGH release. Appropriately called growth hormone releasing agents, these substances would trigger the pituitary to release HGH in small amounts in several pulses during the day, just as it would naturally. Scientists hope that returning this natural rhythm to HGH release would help patients avoid the side effects seen with rHGH use, which involves injecting relatively large amounts of the hormone directly into the bloodstream only once or twice a day. Although these substances are still in development, it's possible they'll be available to the public within ten years or so.

Q: Are there any prescription drugs that boost HGH levels?

Yes. Studies show that four common prescription drugs may boost HGH levels. These drugs include levodopa (Sinemet), used primarily to treat Parkinson's disease; ergoloid mesylates (Hydergine), used to treat Alzheimer's disease; clonidine (Catapres), used to treat high blood pressure; and phenytoin (Dilantin), used to treat seizures in epileptics. Although highly effective in treating disease, these drugs may have serious side effects and risks that far outweigh any relatively minor boost in HGH levels they trigger. In addition, it is unlikely that a physician would prescribe any of these drugs for anti-aging purposes alone.

Q: What about DHEA? Is that as difficult to get or to take as rHGH?

Until recently, DHEA was sold only by prescription. Today, however, you can buy this hormone supplement over the counter at most health food stores and pharmacies. DHEA (short for dehydroepiandrosterone) is a steroid, similar to estrogen, progesterone, and testosterone, produced by the adrenal glands as well as by certain skin and brain cells. During the mid-1980s, scientists discovered that levels of natural DHEA decrease as we age so that by the time we reach our eighties, we have just 15 percent of our youthful levels.

Q: What are the risks and benefits associated with DHEA?

Until the mid-1980s, scientists were unsure that DHEA had any particular role to play in the body, except to provide the raw ingredients required for the body to create estrogen and testosterone, the two main sex hormones. Although scientists remain unsure of exactly how DHEA acts on cells and tissues throughout the body, hundreds of studies show that replacing DHEA as we age engenders many benefits, including

- increasing libido and general vitality
- lowering one's risk for heart disease, and even helping to reverse present cardiovascular damage
- diminishing the effects of Alzheimer's disease and other degenerative brain disorders
- boosting the immune system

However, taking DHEA is not without risk. Like HGH, DHEA stimulates the release of IGF-1, a key mediator in the growth and maintenance of cells throughout the body—including, unfortunately, cancer cells. There exists, then, the potential for too much DHEA to exacerbate the development of certain cancers. The more common, and far less worrisome, side effects of DHEA include

- acne and excessive skin oiliness
- unwanted hair growth in women

- deepening of the voice
- irritability or mood changes
- overstimulation or insomnia
- fatigue or low energy

Many people taking DHEA find that if they simply lower the dose, or switch from taking it every day to taking it every other day, these side effects disappear within just a few days or weeks.

Q: How much DHEA should I take?
Since everyone's body chemistry is a bit different, you'll probably have to experiment a little to find the right dose for you. As discussed, your first step should be to have a physician measure the amount of DHEA that you currently produce by performing a simple blood or saliva test. (Unfortunately, your insurance probably won't cover the cost of these tests, which run from about forty dollars for a saliva test to about ninety dollars for a blood test.) The normal amount of DHEA in a healthy young woman is about 2,600 nanograms per deciliter of plasma and for a man about 3,600. If you're over fifty, your levels are probably much lower than this.

The dose of DHEA supplements suggested to return your DHEA to youthful levels is between 25 and 50 milligrams a day, depending on what your test results show. Many doctors recommend that you start with a low dose (5 to 15 milligrams) and increase slowly to bring your blood or saliva level back up to youthful levels. (Since over-the-counter preparations come in 25-milligram capsules, you may need to find a pharmacist who will compound the right dose for you.) You should monitor your level with monthly lab tests for the first few months, then once or twice a year after that.

Q: Do I have to take DHEA at any particular time of the day?

Unlike HGH and melatonin, both of which the body secretes mostly at night, DHEA levels don't change much during the day. Therefore, you should feel free to take a dose at any time. Because some patients feel a relatively immediate increase in libido following a dose of DHEA, you may want to save your dose for evening.

Q: I'm a woman just entering menopause. Should I take estrogen replacement therapy (ERT)?

That's a question you should answer with the help of your physician. Estrogen, either alone or in combination with the other female hormone called progesterone, has been prescribed for decades to menopausal women. It both helps alleviate the symptoms of menopause (hot flashes, dry skin, mood swings, etc.) caused by the loss of estrogen as well as protects against heart disease and osteoporosis.

Nevertheless, taking ERT engenders some risk as well. The most serious side effect of ERT is its potential to cause breast cancer and endometrial cancer in postmenopausal women. Although the risk of cancer appears to be small in comparison to estrogen's benefits, you should talk to your doctor about how it may relate to your personal medical profile. Other side effects include increased risk of thromboembolic disease (blood clots) and liver disease, as well as fluid retention, abdominal cramps, headaches, and weight gain. These side effects are relatively rare, however, and you may definitely feel, as millions of women do, that the benefits of estrogen outweigh its risks.

Q: What form does estrogen come in?
Estrogen exists in many dosages and forms. Synthetic estrogens are produced in the laboratory using artificial, petroleum-based chemicals. Natural estrogens are identical to those found naturally in the body and are derived from animal sources. In general, the synthetic estrogens are more potent than the natural preparations—in fact, they are used chiefly for the manufacture of the birth control pill, which requires high doses of estrogen in order to suppress ovulation. Almost all women using ERT to relieve menopausal symptoms take natural estrogen.

Of the dozen or so different natural estrogen preparations, the most commonly prescribed brand is a conjugated equine estrogen known as Premarin (in fact, Premarin is now the number one prescription drug in the United States). Premarin comes in both pill form and a vaginal cream that directly treats symptoms of vaginal atrophy. Another common estrogen preparation is estrone, a weak form of estrogen produced both by the ovaries and through the conversion of androgens (male sex hormones) by fat cells. Estrone also comes in pill and cream forms under the brand name Ogen. Estradiol, a more potent form of estrogen, is the main constituent of the pills marketed under the brand name Estrace and in the transdermal patch called Estraderm. Most women who take estrogen therapy must also take progesterone, the other female sex hormone. Progesterone is part of a class of hormones known as progestins. Like estrogen, progestins are either derived from natural animal sources or produced synthetically.

Q: How much estrogen and progesterone should I take?
That's an issue for you and your doctor to decide. Generally speaking, however, most gynecologists would agree that when it comes to hormones, less is best. The more estrogen you take, the greater your risk of developing uncomfortable and potentially dangerous side effects. The average dose is about .625 milligrams per day of estrogen for the first twelve days of the cycle, followed by estrogen and progestin in 5 to 10 milligram doses per day together for about thirteen days. After twenty-five days, you'll stop taking both pills.

Q: How long should I take estrogen?
That's yet another question that we can't answer without knowing more about you. Some scientists believe that a woman should take estrogen only for as long as her symptoms trouble her—maybe for two or three years. After that, she can taper off and then exercise and eat a proper diet in order to protect herself against osteoporosis and heart disease. Other physicians believe that the benefits of taking ERT, which include increased libido and vitality as well as decreased risk of heart disease and osteoporosis, are worth the added risks, even if a woman takes the hormones for the rest of her life.

Q: Do we know as much about the effects of replacing testosterone as we do about estrogen replacement?
Because the loss of testosterone in men occurs much more gradually and to a less dramatic extent

than estrogen does in women, it is only in recent years that scientists have begun to consider the therapeutic effects of replacing the hormone. In both sexes, this steroid hormone helps build muscle, stimulates production of red blood cells, and influences sex drive. Its loss, therefore, has widespread effects on our health.

Testosterone levels begin to decline in men starting at around age fifty, which explains why many men of this age complain of a loss of libido and decrease in energy. A little known fact is that women, too, both produce testosterone normally throughout their lifecycle as well as experience a slight decrease in production after menopause. Some physicians believe that both men and women over fifty can benefit from taking testosterone once symptoms of decreased libido and energy appear.

Q: Should my doctor test for testosterone levels?
Absolutely. In fact, because testosterone is available only by prescription, your doctor will probably insist on monitoring your use of the hormone quite carefully. For women, the normal level of testosterone is about 25 to 100 nanograms per deciliter of plasma; men produce ten to twenty times that amount.

Q: What form does testosterone come in and how much should I take?
Although oral preparations of testosterone are available, taking the hormone in this form may cause liver damage over time, since the liver must metabolize the drug before it can reach the bloodstream. Your doctor may prescribe injections of

about 200 milligrams of testosterone once or twice a month if you're a man with low levels of the hormone. Another alternative is the testosterone patch, which you place on clean, dry skin on the back or buttocks. The patch releases a continuous and steady dose of the hormone. Testosterone is also available in creams, gels, and sublingual pill form—all of which bypass the liver and are thus preferable to oral preparations.

Women who take testosterone will want to take as little of the hormone as possible to achieve the desired result. Many physicians recommend that women take the hormone in cream or gel form that you can rub directly on the skin, often only a few days a month. Talk to your doctor about what form and dose is best for you.

Q: What side effects should I watch for?

Like estrogen in women, testosterone has the potential to trigger the development of cancer of the reproductive organs in men. Of particular concern is prostate cancer, which makes regular digital rectal exams and prostate-specific antigen blood tests all the more important if you decide to take testosterone supplements.

In both men and women, common side effects of testosterone therapy include

- unwanted facial hair
- increased risk of heart disease (testosterone increases LDL, the "bad" cholesterol, while decreasing HDL, the "good" cholesterol)
- deepening of the voice
- irritability and aggression

Q: I've read a lot about melatonin. When should I start taking that hormone?

As is true for the other hormones we've discussed, melatonin production begins to decrease during our twenties until we nearly run out of the hormone in our sixties and seventies. If you're taking the hormone strictly for anti-aging purposes, then, you probably don't need to start taking it until you're over fifty. If you have sleep problems, you can take it for short periods of time at any point in your lifecycle.

Q: How much melatonin should I take, and how often?

That depends on several factors, including your age, your body chemistry, and, most importantly, the reason why you're taking melatonin. If you take it as a treatment for a sleep disorder, for instance, you'll probably want to take about 3 to 6 milligrams about three to four hours before you want to go to sleep. Start with a lower dose, then work your way up if necessary. You should take the supplements until your sleep disorder has been resolved, which can take from just a few days (in the case of jet-lag or stress-related insomnia) to several months (in the case of more deep-seated problems).

If you're using melatonin primarily for its antioxidant, anti-aging effects, you should probably take about 3 milligrams every night after you turn fifty. Always take the hormone in the evening, about three to four hours before you go to sleep.

Q: What will I feel like after taking a melatonin supplement?

Again, that will depend on a number of factors: when you take it, how much you take, and your own particular body chemistry. For most people, a standard dose of 3 milligrams will trigger sleep-related hormonal activities within an hour or two after ingestion. Body temperature will fall and the release of stress hormones that keep the body awake and ready for action will slowly ebb. Soon, feelings of fatigue should set in. These changes should be gradual and feel natural, just as if you were getting sleepy when your body clock is working on its own.

Q: Are there any side effects associated with melatonin?

To date, no long-term side effects and even very few short-term side effects have been noted with melatonin. Among those most often cited are sleepiness during the day, mild stomach cramps, vivid dreaming and/or nightmares, and, paradoxically, increased insomnia. In most cases, these side effects can be alleviated by either changing the amount of the hormone you take, changing the time you take the dose, or both. As is true for the other hormones we've discussed, finding the right dose of melatonin for you may require a little experimentation.

Q: I can see how melatonin might affect me in the short term, but how will I know if melatonin is having any long-term positive effects, like reducing the risk of cancer or turning back my aging clock?

Unfortunately, there's no clear way to measure how well preventive supplements are working. After all, you're attempting only to return your body to its natural state rather than to correct any specific problem and thus the effects tend to be subtle.

Within a few months of taking melatonin on a regular basis, you may find that you are no longer plagued with as many headaches or infections as you once were, or that you feel more rested in the morning and more energetic during the day.

Q: Could melatonin interfere with any other medications I'm taking?

As is true for the other substances we've discussed, melatonin is a hormone and, as such, may have widespread effects on several systems in your body. Although no drug interactions have been reported to date, you should discuss taking any hormonal supplements with your doctor. In the case of melatonin, this warning is especially applicable if you take corticosteroid hormones, such as those used to treat asthma and arthritis, since they boost the levels of hormones that directly react with melatonin. However, taken in moderate doses, the effects of melatonin tend to be quite subtle and therefore, in most cases, should not interfere with most medications.

Q: Are any of these hormones regulated by the FDA?

The Food and Drug Administration supervises the development and marketing of all drugs sold in the United States, both prescription and over the counter. It is a branch of the Department of Health and Human Services, funded annually through the U.S. Congress, and has the authority to regulate foods, drugs, cosmetics, and medical devices that are sold among the states or imported. The FDA is responsible for ensuring that these products are pure and unadulterated, and not misrepresented through false labeling, declarations of ingredients, or net-weight statements. As such, the FDA retains strict control over medications like human growth hormone, estrogen, and testosterone, which are available only by prescription.

The FDA generally does not have jurisdiction over naturally-occurring substances such as vitamins, minerals, enzymes, amino acids, herbs, and certain hormones like DHEA and melatonin—as long as health claims are not made on their behalf by their manufacturers. In other words, the laboratories that produce and sell DHEA cannot proclaim that you'll prevent heart disease by taking the supplements or even that melatonin can treat sleep disorders. If they did, the FDA could declare them to be drugs subject to the same strict rules and regulations as prescription and over-the-counter pharmaceuticals.

Q: If substances aren't regulated, how can I be sure what I'm taking is safe?

You're right to be concerned about the quality of the supplements you take—and that's why you must never purchase human growth hormone on the black market. As discussed in Chapter One, some stores of rHGH may be contaminated with a virus that causes a fatal brain disease called Creutzfeldt-Jakob disease. The role of the FDA is to ensure that powerful substances like HGH are manufactured safely.

It's important to note, however, that the safety record for vitamin, mineral, and other natural supplements—including melatonin and DHEA, has been consistently outstanding. According to summaries from the nation's poison-control centers, only one death was associated with the use of a nutritional supplement from 1983 to 1990, and that was due to the overuse of niacin by a mentally unstable individual. Prescription drug reactions or interactions, on the other hand, caused approximately 130,000 deaths *every year* during the same period. Nevertheless, it's important that you remain responsible about your health and read up on the status of the supplements you take through news reports and frank discussions with your physician.

Q: If I decide not to take hormones, what can I do to stay as young and vital as possible for as long as possible?
The habits that will keep you healthy—eating right, exercising regularly, getting enough sleep, and reducing stress—are the very ones that boost hormone levels naturally and help you stay young. In Chapter Nine, we discuss those healthy habits and help you make them part of your daily life.

NINE

Living Well Longer

Q: What are the things I can do—apart from taking supplements—to stay healthy and vital as I age?

In Chapter Three, we listed seven habits that studies show increase longevity: consuming only moderate amounts of alcohol; eating breakfast on a regular basis; maintaining a healthy weight; avoiding fat-laden, sugary snacks; getting regular exercise; enjoying seven to eight hours of sleep a night; and never smoking cigarettes. Without question, these habits will help you stay as young and healthy as possible for as long as possible. One reason is that by eating well, exercising regularly, and getting plenty of sleep, you allow your body to function at its optimum capacity, which also means producing sufficient amounts of hormones like HGH for as long as possible.

If you smoke, stop as soon as you can. Talk to your physician about using one of the nicotine patches now available over the counter, or by prescription, if you need chemical support during this difficult process. Contact the American Lung Association or the American Heart Association, both of which offer stop smoking programs that can help you kick a habit that provides no benefits to outweigh its dreadful risks.

When it comes to drinking alcohol, moderation is the key. In fact, if you keep your intake of alcohol to one drink a day, you may actually derive some health benefits. Studies show that alcohol may significantly reduce your risk of developing heart disease and hypertension. However, you must weigh that benefit against the many known risks of alcohol consumption: Without question, the more alcohol you consume, the more damage you can do to the cardiovascular system and the rest of your body. Indeed, after tobacco, alcohol abuse is the leading cause of premature death in the United States and is associated with the loss of more than 100,000 lives annually. Studies show that even moderate drinkers (those who drink more than one or two drinks a day on a regular basis) may run an increased risk of breast cancer and heavy drinkers have high rates of liver disease. Alcoholics who continue to drink decrease their life expectancy by ten to fifteen years.

The other habits that help ensure a longer, healthier life relate to diet, exercise, and sleep. Eating breakfast every morning, for instance, helps the body's metabolism kick in, allowing the body and the brain to function at their highest capacity and helping the body burn energy more efficiently. The benefits of avoiding sugary, fatty foods are self-evident: In addition to helping us keep a healthy body weight, this habit encourages us to eat healthier foods. Exercise provides more benefits than we can count, as we'll discuss later in the chapter.

We add one other habit that the previous study overlooked: stress reduction. Although stress is hard to measure, and its effects difficult to calculate with any consistency, scientists are now convinced that the way an individual handles stressful situations has a direct impact on his health and longevity. Specifically, stress may interfere with the way hormones like HGH, DHEA, and melatonin function, as well as cause damage to the organs of the endocrine system. A side effect of stress reduction, therefore, may be an increase in the amount and effect of anti-aging hormones your body produces naturally.

Q: It's all so confusing. I never know what to believe when it comes to a "healthy lifestyle."

You're not alone in your confusion and frustration. For many Americans, the whole subject of diet, exercise, and stress reduction has become fraught with tedium, frustration, and often anxiety. A newspaper headline proclaims the virtues of using margarine instead of butter, then—practically the next day—the same newspaper runs a story about the dangers of trans-fatty acids, an essential ingredient in margarine. One report claims that exercise is beneficial only if you exercise at your target heart rate three times a week for thirty minutes; another maintains that fitness can be achieved simply by increasing your daily physical activity an extra fifteen or twenty minutes.

Because new information emerges every day that changes yesterday's "rules," it's safe to say that no one really knows *exactly* what it takes to stay healthy and fit. Nevertheless, there are some general prescriptions to follow that will help you along the right path, and we discuss those guidelines in this chapter. In the meantime, it's important to stress that the purpose of changing your daily habits is not to add frustration and anxiety to your life. Instead, your goal should be to add these three elements to your lifestyle.

- **Balance**. One definition of the word *balance* is "a state of stability, as of the body or emotions;" another refers to *balance* as a "state of harmony." Is your life in balance? Or are you so busy you feel you don't have time to exercise or eat right? Do you pay more attention to your physical than emotional health? Or do you spend more time on other people than on yourself? Restoring balance to your life will help your body function in the smooth and rhythmic way in which it was designed.

- **Structure**. Today's fast-paced world leaves many of us trying to do too much in too little time. The result is that tend not to eat, exercise, sleep, or relax on a regular basis. We become distracted and overwhelmed, grabbing fast-food lunches at three in the afternoon, putting off exercise until the weekend (when we tend to overdo it), and sleeping only in fits and starts during the night. By adding structure to our lives—by scheduling specific times for our meals and exercise routines—we help our bodies establish a natural rhythm once again. This rhythm directly impacts on the release of hormones throughout the day and night.

- **Satisfaction**. In the end, what matters most to your ability and, perhaps, even your desire for health and longevity is that the life you live is a satisfying and fulfilling one. Again, establishing a healthy lifestyle should never be a chore to be dreaded or despised, but instead an opportunity to provide your body and soul with the raw ingredients needed to thrive and flourish. If your diet leaves something to be desired, create an eating plan that not only bolsters your protein intake in order to help your body maintain muscle, but also includes foods you enjoy eating. If you're looking to start an exercise program, choose physical activities that bring you pleasure as well as burn calories and trigger hormone release.

Q: When it comes to stimulating the release of natural human growth hormone, what's the most important food for me to eat?

In order to manufacture human growth hormone as well as to build and maintain muscle and bone tissue, your body needs adequate amounts of protein. Next to water, protein is more plentiful than any other substance in the body—all the tissues, bones, and nerves are made up mostly of proteins. Some hormones are proteins as well, notably the thyroid hormone and insulin, which control a variety of body functions such as growth, sexual development, and rate of metabolism.

As discussed in Chapter Two, the process of digestion breaks down protein into simpler units called amino acids. Amino acids are necessary for the synthesis of complete body proteins and many other tissue components. They are the units from which proteins are constructed, and some appear to act as potent releasers of human growth hormone.

The body requires approximately twenty-two amino acids in specific patterns to make human protein. The body itself can make all but nine of these amino acids. The nine the body cannot produce are called "essential amino acids" because you must receive them in your diet. The nine essential amino acids are lysine, methionine, threonine, tryptophan, leucine, isoleucine, valine, phenylalanine, and histidine. Among the foods highest in protein—and thus in essential amino acids—are meat, fish, and poultry, soybean products, eggs, milk and milk products, and whole grains.

Q: How much protein do I need to eat every day?

The minimum daily protein requirement—the smallest amino acid intake that can maintain optimum growth and good health in an adult—is 0.8 grams of high-quality protein per kilogram of average body weight for height. The National Research Council recommends about twice that, or 0.45 grams of protein per day for each pound of body weight. To figure out your requirements, simply divide your body weight by two: The resulting figure will be the grams of protein you need to eat every day. For example, if you're a healthy woman who weighs 130 pounds, you require about 65 grams of protein every day. Of course, protein requirements differ according to your nutritional status, body size, and activity level. In general, about 15 percent of your daily intake of calories should consist of protein.

Q: I'm a vegetarian. Can I meet my need for protein and amino acids without eating meat, fish, or poultry?

Yes, but you have to be particularly careful about the foods you choose. In order for the body to properly synthesize protein, all the essential amino acids must be present at the same time and in the proper proportions. If just one essential amino acid is low or missing, protein synthesis will fall to a very low level or stop altogether.

We call foods that contain all the essential amino acids "complete proteins." Meats and dairy products are complete protein foods, while most vegetables and fruits are incomplete proteins. To obtain a complete meal from incomplete protein foods, you must combine these foods carefully so that you balance foods missing one or more essential amino acids with foods that contain those substances.

You can design your diet to have sufficient complete protein content. If you combine beans with corn, nuts, rice, seeds, or wheat, for instance, you'll be eating a complete protein. When made with whole wheat bread, even a peanut butter and jelly sandwich qualifies as a complete protein.

If your diet allows you to eat eggs, you'll find it easy to meet your protein needs. Egg is a high-quality protein which, although high in cholesterol, is rich not only in protein but in other nutrients as well. Soy products such as soy beans and tofu are complete plant proteins that have all the benefits of red meat but without the animal fat that can be so dangerous to our health if eaten in excess.

Q: What about the rest of my diet? How do I create some balance there?

The United States Dietary Association (USDA) developed a way for the average American to plan a healthy daily diet. According to this plan, each day you should eat the following.

- 6 to 11 servings of complex carbohydrates (bread, cereal, rice, and pasta)
- 2 to 4 servings of fruit
- 3 to 5 servings of vegetables
- 2 to 3 servings (3–5 ounces each) of lean protein (meat, fish, eggs, beans, nuts)
- 2 to 3 servings of low-fat dairy products (milk, yogurt, cheese)
- Limited fat and sugar

If you're concerned about your diet, it's best to seek advice from a nutritionist or other health professional. He or she can help you adapt these general guidelines to fit your needs in terms of calorie intake and specific nutrient requirements.

Q: One of the reasons I'm thinking about using rHGH is because it's supposed to help me lose weight. If I don't take rHGH to reduce my body fat, how else can I drop the pounds? I hear constantly that "diets don't work."

You're not alone in your confusion. The American public has been receiving mixed messages about dieting and weight maintenance for too long. The fact is, diets *don't* work—at least not the ones that strictly limit the type of food you eat on a daily basis. Instead, it is suggested that people who lose weight successfully and permanently do so by limiting the *amount* of food they eat, choosing lower fat foods, and by increasing their activity levels. If you want to gain some control and balance over your diet, start by making small changes, one by one. Tomorrow, for instance, trade your jelly doughnut for a bowl of oatmeal with raisins at breakfast, drink two glasses of water instead of a soda for lunch, and grill a piece of fish instead of a steak for dinner. Slowly but surely, you're sure to see the pounds start to drop off and your energy levels increase.

Q: Are there particular exercises I should do that will increase levels of HGH and other hormones?
Don't worry about what kind of exercise to get— just get out there and move! Although weight-bearing exercises (such as lifting weights or jogging) work more efficiently with HGH to build muscle and maintain bone density than aerobic exercise, any form of exercise will help you feel better, look better, and improve your health. Furthermore, any exercise will trigger the endocrine organs to increase their secretion of HGH, DHEA, testosterone, estrogen, and melatonin.

Q: I know I should exercise, but never seem to have the discipline to keep up the routine. Any suggestions?
Again, we can look to the three watchwords for a

healthy life: balance, structure, and satisfaction. First, balance your energy. Part of your resistance to exercise may come from the idea that you must devote your life to getting and staying in shape. In fact, exercise should fit into your life in a natural, balanced way.

For instance, exercise is not an all-or-nothing proposition. You may think that you must run every single morning or make every evening's aerobics class at the gym. If you can't manage these demands, you're likely to give up. But this is an outdated mode of thinking. Exercise can consist of a brisk walk to the market, an afternoon of gardening, even an hour of vigorous vacuuming—any activity that requires your body to move and your mind to take a backseat to the task at hand. You should also try to set realistic goals and avoid injury by overexercising.

Your ultimate goal is to make exercise a habit rather than an occasional, dreaded assignment. To do so, you should schedule convenient times and places to exercise: If you join a health club that is open only during hours you are at work, for instance, then obviously you're setting yourself up to fail. Finally, you need to find activities that stimulate and satisfy you. Indeed, nothing will sabotage an exercise program faster than boredom and/or frustration. Perhaps the most important element in the design of your exercise program is choosing activities you will enjoy over the long haul. If your primary purpose in exercising is to stimulate HGH release, you'll want to include some weight-bearing exercises, however, any exercise done on a regular basis will boost your energy levels and improve your general health.

Q: I find that if I exercise, I sleep better. Will that happen even as I get older?

A recent study showed that people over age eighty-seven who performed forty-five minutes of aerobic exercise three times a week were getting 33 percent more deep sleep than before the study began and were secreting 30 percent more HGH during their sleep time than those who did not exercise. It's important to understand, however, that if you do make exercise a regular part of your life you will not only *want* to sleep more, you'll *need* to sleep more. Your body will need sleep—and the hormones triggered during sleep—in order to repair the stress exercise puts on the muscles and bones during a hard workout.

Q: Why is sleep so important?

We fall into sleep for two main reasons: to conserve energy and to recuperate from the previous day's activities. First, we conserve energy when we sleep because our metabolic rate—the rate at which our bodies use energy to function—is reduced by at least 25 percent over daytime levels. Oxygen consumption, heart rate, and body temperature all decline during the first few hours of sleep and reach an all-time low about an hour before we wake up. In essence, then, sleep allows the body to maintain homeostasis—a relative stable internal environment.

As for the recuperative and restorative powers of sleep, the body releases more than half of its daily output of HGH, which helps to repair and maintain tissue during the night and throughout the following day. In addition, sleep allows the whole body, including the central nervous system, to repair itself. Cortisol and other glucocorticoids, known to stimulate heart and respiration rates, fall to their lowest levels during sleep.

Many sleep experts believe, however, that it is the brain and not the body that benefits most from sleep. Indeed, studies show that sleep deprivation results in far more psychological than physical deficits—our concentration falters and our mood darkens far more quickly than our bodies experience any lasting physical impairments. In fact, people with insomnia or other sleep problems are more likely to develop psychiatric illnesses than their sleep-sated peers.

The truth is, however, that sleep researchers still don't know with any certainty the specific biological functions of sleep. Indeed, people kept awake in experiments for as long as eleven days straight have experienced no discernible physiological damage and only minor changes in circadian hormonal rhythms. On the other hand, no one can deny that a lack of sleep makes us at least feel bad—our mental performance and memory tend to suffer, we become irritable and/or anxious, and our physical reflexes falter, however temporarily.

Q: How do sleep patterns change as we age?

Infants sleep roughly twice as much as adults—up to eighteen hours a day—but by age ten or so, most kids are down to about ten hours a night. Although some teens sleep more than their parents think possible, most develop sleep patterns that approximate those of adults.

The next dramatic shift appears in the elderly. By the time we reach our sixties and seventies, our sleep often becomes fragmented and disorganized.

In fact, although older adults tend to spend about one hour more per night in bed, the biological rhythms that control sleep don't hold together as well. Dr. Charles Czeisler of Harvard's Brigham and Women's Hospital and an expert in circadian rhythms has found through work at his lab that sleep cycles become shallower and more rapid in older people for a number of reasons. (In fact, the average sixty-year-old awakens more than twenty-two times a night, while a younger person in his or her twenties awakens just ten times.)

One reason we tend to sleep less as we age is that the pineal gland—a tiny organ in the center of the brain—begins to shrink, thereby inhibiting the release of sleep's most ardent promoter: melatonin. Responsible for setting the physiological stage for sleep, melatonin's loss makes it difficult both to fall asleep in the first place and stay asleep once it does.

Second, lifestyle changes that occur with aging also may have a significant impact on sleep patterns. Many older adults fail to get enough exercise—partly because their muscles and bones have lost mass and strength due to the loss of HGH—and thus never really tire themselves out in a physical sense. Others develop low-grade depression and, though they feel lethargic, are too anxious and upset to sleep well.

Q: What happens during sleep?
Sleep is composed of two distinct physiological states—rapid eye movement (REM) and non-REM—as different from each other in some ways as each one is from wakefulness. We pass through about four to six non-REM/REM sleep cycles every night. Let's look at each stage separately.

- *Non-REM sleep* consists of four different phases: *Phase One* is a light, drowsy sleep representing the transition from wakefulness to sleep and then, later, from sleep to wakefulness. *Phase Two* is the first real stage of sleep, when your brain and body descend into unconsciousness. *Phases Three and Four* are known collectively as "slow wave sleep" or "delta sleep." It is during this stage that body recovery is thought to occur. Blood, filled with growth hormone and nutrients, flows directly to muscles and organs, and the bone marrow increases its production of red blood cells.

- *REM sleep*, on the other hand, involves a dramatic increase of blood flow to the brain: as much as a quarter of all blood circulating during REM sleep travels through the brain. Presumably, this increased blood flow is needed to support the many activities taking place in the brain during this period. First of all, we dream during REM sleep, and as we dream, our heart rate and blood pressure rises, metabolism speeds up, and our breathing gets faster and faster and more irregular—all triggered by signals from the hypothalamus and other endocrine organs. Second, sleep researchers believe that short- and long-term memory are stored and other cognitive functions processed during REM sleep.

As people fall asleep, they progress through the non-REM sleep stages and, then, about ninety minutes later, they have their first episode of REM sleep. As the night progresses, the episodes of non-REM sleep become shorter and those of REM sleep longer. Most slow wave sleep—and thus HGH release—occurs during the first third, and most REM sleep occurs during the last third, of the sleep cycle.

Q: How long does the average person sleep?
As discussed, sleep patterns are often age-dependent. However, the majority of people (about 84 percent) sleep about 7.5 hours every night. But 15 percent of Americans get only 5.5 to 6.5 hours, while another 15 percent manage a full 8.5 hours. About one in 100 people seem to thrive on only 5.5 hours, while another one in 100 seem to need 10.5 hours or more.

Q: How do I know if I'm getting enough sleep?
A good rule of thumb is that if you wake up feeling rested and ready for action, you've slept enough. But because of the generally hectic pace of modern life, and the amount of external stimuli you probably receive on a day-to-day basis, you may no longer feel any connection to your physical body or know how to listen to the signals about the need for rest and replenishment your brain may be sending you. This lack of connection—and lack of sleep—may be why stress-related, degenerative diseases like heart disease and arthritis are so prevalent.

Q: Is how long you sleep related to how tired you are or how long you've been awake?
The length of time a person sleeps is related more closely to body temperature rhythms and bedtime

than to how long the person has been awake or even how exhausted he or she feels. In one study, even after being awake more than twenty hours, people free of time cues slept twice as long when they went to bed when their temperatures were at their highest (in the early evening) than when they were at their lowest (early in the morning).

Q: Can we repair the damage done by lack of sleep?

A single night's sleep, one in which you let yourself sleep until your own body clock wakes you, is usually enough for you to regain about 90 percent of the mental acuity sleep deprivation caused you to lose. A second full night of sleep restores the remaining 10 percent. In sleep experiments, people kept awake for three days then slept for ten hours for two consecutive nights and then, by the third night, returned to their normal sleep patterns.

Q: What exactly is stress and how do I measure how much I'm affected by it?

That's a good question, one that scientists still can't answer with any certainty. The difficulty many physicians have in assessing stress is threefold: First, except for extreme situations, like the death of a loved one or the threat of imminent physical harm, a clear definition of stress is unavailable. Everything that occurs in your life or exists in the atmosphere is technically a stressor because it affects you in some way. If it is very hot out, for instance, your body will adjust to the increased temperature by cooling the skin with perspiration. In this instance, heat is a stressor because it spurs the body into action. If you find out you've just become a grandfather for the first time, the excitement the event stimulates may make your heart beat faster, your muscles tense up, your palms sweat. Despite its positive impact, then, the news about the birth of your grandchild is a stressor because it forces a physiologic reaction to occur.

Second, clearly, not everyone responds to stress in the same way. Some people become outwardly aggravated over the slightest mishap while others never blink an eye even when disaster occurs. At the same time, the outwardly calm person may be actually seething inside, perhaps negatively affecting his or her physiology even more than the person who expresses anger and frustration in a more open way.

Third, since each of us has a different idea of what stress is, you begin to see how complex this issue is. Indeed, stressors vary quite dramatically from person to person. For some, a day spent lying on a beach is completely relaxing, while for others such forced recreation is sheer torture. It is how you as an individual perceive an event that determines how your body reacts to it.

Q: But what *is* stress, and what does it do to the body?

Stress represents one of your body's best methods of self-preservation. When its internal balance is threatened in any way, it mobilizes immediately, preparing you either to battle the impending danger or to flee from it. Called the "fight-or-flight" response, this automatic reaction primarily involves the autonomic nervous system, the part of the nervous system responsible for regulating bodily functions such as the heartbeat, intestinal movements, muscular contraction, and hormonal secretion. It is divided into two parts that work to balance these activities: The sympathetic nervous system speeds up heart rate, raises blood pressure, and tenses muscles during times of physical and emotional stress, while the parasympathetic nervous system works to slow these processes down when the body perceives that stress has passed.

During times of stress, then, your entire metabolic rate increases, leading to symptoms such as high blood pressure, neck and backaches, dizziness, diarrhea, fatigue, insomnia, sexual problems, and frequent illness due to an immune system that is also overstimulated. In addition, stress upsets the normal balance of hormones in the body, releasing some hormones in quantity while reducing the secretion of others.

Specifically, the two most powerful stress-related hormones are norepinephrine and epinephrine, both produced by the adrenal glands. These hormones stimulate the sympathetic nervous system to raise blood pressure and heart rate, to make you breathe in more oxygen, and to cause your muscles to tense up.

Q: What methods help reduce stress?

Fortunately, there are several ways to improve the way you cope with the stressors in your life. Perhaps the simplest method involves deep breathing. Have you noticed that whenever you're distressed, your breathing becomes rapid and shallow, but when you're relaxed, you breathe more slowly and regularly? Or that when you're tired, a deep breath makes you more alert? By learning to breathe deeply and in a rhythmic fashion, you can bring your body and mind back into balance whenever you're faced with a stressor. Here's one exercise you can try.

1. Sit on the floor in any comfortable position. Make sure your back is straight (neither arched or leaning forward), your head is erect and facing forward, and your arms are relaxed, with your hands resting on your thighs or on the floor.

2. Close your eyes and concentrate only on your breathing. Leave behind the worries or joys of the day and think only of this moment in time, when you are feeling the energy and power of your breathing.

3. Think of your lungs as consisting of three parts: a lower space located in your stomach, the middle part near your diaphragm (just beneath your rib cage), and the upper space of your chest.

4. As you breathe in through your nostrils, picture the lower space filling first. Allow your stomach to expand as air enters the space. Then visualize your middle space filling with energy, light, and air, and feel your waistline expand. Finally, feel your chest and your upper back open up as air enters the area. The inhalation should take about five seconds.

5. When your lungs feel comfortably full, stop the movement and the intake of air.

6. Exhale through your nostrils in a controlled, smooth, continuous stream. Feel

your chest, your middle, and your stomach gently contract.

7. Make about four complete inhalations-exhalations in a minute, resting about two or three seconds between breaths. Rest for twenty seconds or so, then repeat the process until you feel more relaxed and in control.

8. If you like, add a self-affirmation to your relaxation session by saying to yourself, "I am in control," "I am relaxed," "I will concentrate," or another positive motivating phrase every time you exhale.

This deep breathing exercise is just one way you can reduce the stress in your life. Yoga and meditation are others. In the Resources, Reading, and Reference section on page 204, you'll find information about stress-reduction methods and how you can build them into your daily life.

Q: Even if we follow all of these recommendations, even if hormones like HGH really work, will aging ever become a thing of the past?

Probably not. As far as we know, every living thing in the universe has a life cycle that includes aging and then, finally, death. And the truth is, we probably wouldn't—or couldn't—have it any other way. If humans never became old and died, the planet would not be able to sustain life of any kind for very long.

Instead, we should strive to live with as much health, vitality, and vigor as possible, every day, for as long as possible. One way to do just that is to follow at least some of the recommendations laid out in this chapter. Indeed, quick fixes such as injections of rHGH or plastic surgery to take away wrinkles, or liposuction to remove fat may be less important to our health and longevity—at least for now—than adding those habits that can improve the quality of your life every day.

＄

Resources, Reading, and References

As we stand on the brink of the twenty-first century, the science of anti-aging represents a new frontier in medicine and, indeed, in the history of human development. The very fact that you picked up and read this book means that you're an interested reader, a wise consumer, and, perhaps, a potential user of new medications and supplements. For that reason, we provide you with a list of agencies and associations concerned with aging and health, a reading list of books offering different perspectives on a variety of topics, and a compilation of research material, largely from the medical press, used to write this book. We hope you avail yourself of these avenues of information as you go forward on your journey of health and vitality.

Resources and Reading

Aging and Anti-Aging

Our interest in aging or, rather, in remaining as young and vital as long as possible, has never been greater. The following reading list offers titles about specific anti-aging therapies as well as about the psychosocial issues surrounding aging. The three associations listed here can provide you with a wealth of information about longevity, including lists of pharmacies and physicians in your area that can help you decide which anti-aging hormone replacement therapies are right for you.

Life Extension Foundation
995 Southwest 24th Street
Fort Lauderdale, FL 33315
800-841-5433
Internet Access: http://www.lef.org

The American Academy of Anti-Aging Medicine
401 N. Michigan Avenue
Chicago, IL 60611
312-527-6733
Internet Access: http://www.worldhealth.net

Longevity Institute International
87 Valley Road
Montclair, NJ 07042
201-746-3533
FAX 201-746-4385

For other information about aging, contact

American Association of Retired Persons (AARP)
601 E Street NW
Washington, DC 20049
202-434-2277
Internet Access: http.//www.aarp.org

Reading List for Aging and Anti-Aging

Case, R. M. and Waterhouse, J. M. *Human Physiology: Age, Stress, and the Environment*. New York: Oxford University Press, 1994.

Cherniske, Stephen, M.S. *The DHEA Breakthrough*. New York: Random House, 1996.

Fossel, Michael, Ph.D., M.D. *Reversing Human Aging*. New York: William Morrow and Company, Inc., 1996.

Friedan, Betty. *The Fountain of Age*. New York: Simon & Schuster, 1993.

LeVert, Suzanne. *Melatonin: The Anti-Aging Hormone*. New York: Avon Books, 1995.

Milunsky, Aubrey, M.D. *Heredity and Your Family's Health*. Baltimore: Johns Hopkins Press, 1992.

Regelson, William, M.D. and Colman, Carol. *The Melatonin Miracle*. New York: Simon & Schuster, 1995.

Regelson, William, M.D. and Colman, Carol. *The Superhormone Promise*. New York: Simon & Schuster, 1996.

Cardiovascular Disease

Heart disease and related conditions like high blood pressure and high cholesterol levels affect more Americans—and prevent more of us from aging well—than any other single disease or condition. For more information about heart disease and how to prevent or treat it, contact

The American Heart Association
7272 Greenville Avenue
Dallas, TX 75231
214-373-6300
Internet Access: http://www.amhrt.org

Reading List for Cardiovascular Disease

Cooper, Kenneth. *Overcoming Hypertension.* New York: Bantam, 1990.

Ornish, Dean. *Reversing Heart Disease.* New York: Ballantine, 1990.

Rothfeld, Glenn S. and LeVert, Suzanne. *Natural Medicine for Heart Disease.* Emmaus, PA: Rodale Press, 1996.

Zaret, Barry L., M.D., ed. *Yale University School of Medicine Heart Book.* New York: Hearst Books, 1992.

Diabetes

Experts estimate that at least fifteen million Americans, most of them older, suffer from this potentially life-threatening disease. As discussed, the relationship between HGH, insulin, and diabetes is still under investigation. The American Diabetes Association will provide you with the most up-to-date information about all aspects of diabetes and diabetes care, as well as information about current and future research.

American Diabetes Association
1660 Duke Street
Alexandria, VA 22314
703-549-1500
Internet Access: http://www.diabetes.org

Reading List for Diabetes

Biermann, June and Toohey, Barbara. *The Diabetic's Book.* Los Angeles: Jeremy P. Tarcher, 1990.

Bernstein, Gerald and LeVert, Suzanne. *If It Runs in Your Family: Diabetes.* New York: Bantam, 1994.

Growth Disorders

Growth disorders affect about five hundred thousand children, many of whom now receive rHGH as treatment. If you need more information about hypopituitarism, achondroplasia, Turner's syndrome, or other conditions known to disrupt growth, contact

Human Growth Foundation
7777 Leesburg Pike, Suite 202S
Falls Church, VA 22043
800-451-6434

Little People's Research Fund, Inc.
80 Sister Pierre Drive
Towson, MD 21204
800-232-5773

Magic Foundation for Children's Growth
1327 N. Harlem Avenue
Oak Park, IL 60302
800-3 MAGIC 3
Internet Access: magic@nettap.com

Menopause

More than 1.5 million women pass through menopause every year, and as they do, they face new challenges to maintaining their health and increasing their vitality. If you're a woman entering menopause, organizations like the local YWCA, American Red Cross, family or community service agencies, department of health, or Planned Parenthood offer support for the many changes—medical and psychosocial—taking place in your life. If you'd like more information about menopause or would like to join or start a support group for menopausal women, write or call this clearinghouse.

North American Menopause Society
c/o Cleveland Menopause Clinic
11100 Euclid Avenue
Cleveland, OH 44106
216-844-8748
Internet Access: http:// www.menopause.org
E-mail nams@apk.net

Reading List for Menopause

Greer, Germaine. *The Change*. New York: Knopf, 1993.

Jovanovic, Lois, M.D. and LeVert, Suzanne. *A Woman Doctor's Guide to Menopause*. New York: Hyperion, 1993.

Sheehy, Gail. *The Pause*. New York: Random House, 1992.

Osteoporosis

Brittle bones caused by the body's inability to repair and maintain bone strength and density as it ages is a leading cause of disability and death, especially in older American women after they lose estrogen. For more information about osteoporosis and how to prevent or limit it, write or call

National Osteoporosis Foundation
1150 17th St NW
Washington, DC 20036
202-223-2226
Internet Access: http://www.nof.org

Reading List for Osteoporosis

Cooper, Kenneth. *Preventing Osteoporosis*. New York: Bantam, 1989.

Nachtigall, Lila, M.D. and Heilman, Joan. *Estrogen*. New York: Harper-Collins, 1991.

Sherrer, Yvonne R., M.D. and Levinson, Robin K. *A Woman Doctor's Guide to Osteoporosis*. New York: Hyperion, 1995.

Sleep Disorders

Difficulties in getting to and staying asleep may create or exacerbate health problems at any age, but seem to be particularly intractable the older we get. In some cases, sleep problems may result from stress that you might learn to better control through biofeedback and meditation techniques. Other cases may benefit from treatment with melatonin, as discussed in Chapter Three, and you can refer to the list of anti-aging associations for more information about melatonin. Still other sleep disorders require medical attention from trained sleep disorder professionals. Call, write, or visit the homepage of this national organization for more information.

American Sleep Disorders Association
1610 14th Street
Rochester, MN 55901
507-287-6006
Internet Access: http://www.wisc.edu/asda/

Reading List for Sleep Disorders

Hauri, Peter, Ph.D and Linde, Shirley, Ph.D. *No More Sleepless Nights*. New York: John Wiley & Sons, 1990.

References

Chapter One

Amato, G., Carella, C., Fazio, S., et al. "Body composition, bone metabolism, and heart structure and function in growth hormone-deficient adults before and after GH replacement therapy at low doses." *Journal of Clinical Endocrinology and Metabolism*, 77: 1671–1676, 1993.

Beaven, Colin. "Pssst—wanna feel young again?" *Esquire*, pg. 50, June, 1996.

Bellino, F.L., Daynes, R.A., Hornsby, P. J., Lavrin, D.H., and Nester, J., eds. "Dehydroepiandrosterone (DHEA and Aging)." *Annals of New York Academy of Sciences*, Vol. 74, 1995.

Brody, Jane. "Restoring Ebbing Hormones May Slow Aging." *New York Times*, pg. B5–6, July 18, 1995.

Cancer Prevention Coalition, press conference. "New study warns of risks of breast and colon cancer from rBGH milk." National Press Club, Washington, D.C., January 23, 1996.

Carey, B. and Lee, K. "The Slumber Solution." *Health*, pg. 70 (8), July-August, 1996.

Cohn, L., Feller, A.G., Draper, M.W., et al. "Carpal tunnel syndrome and gynaecomastia during growth hormone treatment of elderly men with low circulating IGF-I concentrations." *Clinical Endocrinology*, 39 (4): 417–25, 1993.

Geler, A. "Insulin-like growth factor-1 inhibits cell death induced by anticancer drugs in the MCF-7 cells: involvement of growth factors in drug resistance." *Cancer Investigation*, 13(5): 480–486, 1995.

Holloway, L., Butterfield, G., Hintz, R.L., et al. "Effects of recombinant human growth hormone on metabolic indices, body composition, and bone turnover in healthy elderly women." *Journal of Clinical Endocrinology and Metabolism*, 79(2): 470–479, 1994.

Kalimi, M., and Regelson, W., eds. *The Biologic Role of Dehydroepiandrosterone (DHEA).* Berlin-New York: Walter de Gruyter, 1990.

Kotulak, R.A. and Gorner, P. " 'Youth Drugs' Give Old Dreams New Life." *Chicago Tribune*, pg. 35, December 9, 1991.

LeRoith, D. and Clemmons, D. "Insulin-like growth factors in health and disease." *Annals of Internal Medicine.* 116(10): 854–862, 1992.

Orme, S.M., et al. "Cancer incidence and mortality in acromegaly: a retrospective cohort study." *Journal of Epidemiology*, Supplement 0C22, 1996.

Regelson, W. and Kalimi, M. "Dehydroepiandrosterone (DHEA)—The Multifunctional Steroid." *Annals of the New York Academy of Sciences*, 564–575, 1994.

Stirling, H.F. and Kelner, C.J. "Who needs growth hormone?" *Journal of Research in Social Medicine*, 87:497–498, 1994.

Chapter Two

Allanson, J.E. and Hall, J.G. "Obstetric and gynecological problems in women with chondrodystrophies." *Obstetrics and Gynecology*, 67:74–78, 1986.

Bengsston, B.A., Ed'en, S., Lonn, L., et al. "Treatment of adults with growth hormone deficiency with recombinant human GH." *Journal of Clinical Endocrinology and Metabolism*, 76(2):309–317, 1993.

Bouillon, R. "Growth Hormone and Bone." *Hor-*

mone Research, 36(Supplement 1): 49–55, 1991.

Burman, P. "Quality of life in adults with growth hormone (GH) deficiency: Response to treatment with recombinant human GH in a placebo-controlled 21-month trial." *Journal of Clinical Endocrine Metabolism*, 80 (12):3585–3590, 1995.

Cuneo, R.C. and Salomon, F. "The growth hormone deficiency syndrome in adults." *Clinical Endocrinology*, 37: 387–397, 1992.

Dandona, P., Thusu, K., Cook, S., et al. "Oxidative damage to DNA in diabetes mellitus." *The Lancet*, 237: 444–446, 1996.

DeBoer, H., Block, G.J., and Van der Veen, E.A. "Clinical aspects of growth hormone deficiency in adults." *Endocrine Review*, 16: 63–86, 1995.

Jorgensen, J., Vahl, N., Tansen, T., et al. "Influence of growth hormone and androgens on body composition in adults." *Hormone Research*, 45: 94–98, 1996.

Lane, R., Gunczler, P., Paoli, M., and Weisinger, J.R. "Bone mineral density of prepubertal age girls with Turner's syndrome while on growth hormone therapy." *Hormone Research*, 44: 168–171, 1995.

O'Sullivan, A.J., Kelly, J.J., Hoffman, D.M., et al. "Body composition and energy expenditure in acromegaly." *Journal of Clinical Endocrinology and Metabolism*, 78(2): 381–386, 1994.

Chapter Three
Abbasi, A.A., Drinka, P.J., Mattson, D.E., et al. "Low circulating levels of insulin-like growth factors and testosterone in chronically institutionalized elderly men." *Journal of the American Geriatrics Society*, 41(9): 975–982, 1993.

Armstrong, S.M. and Redman, J.R. "Melatonin: A Chronobiotic with Anti-Aging Properties?" *Medical Hypotheses*, 34:300–309, August, 1990.

Belanger, A., Candas, B., Dupont, A., et al. "Changes in serum concentration of conjugated and unconjugated steriods in 40- to 80-year-old men." *Journal of Clinical Endocrinology and Metabolism*, 79:1086–1090, 1994.

Davidson, J.M., Chen, J.J., Crapo, L., and Gray, G. "Hormonal changes and sexual functioning in aging men." *Journal of Clinical Endocrinology and Metabolism*, 57:71–79, 1983.

Foreman, Judy. "Making Age Obsolete." *Boston Globe*, September 27, 1992.

Papadeakis, M., and Grady, D. "Growth hormone replacement in healthy older men improves body composition but not functional ability." *Annals of Internal Medicine*, 124:708–716, 1996.

Rowe, John and Kahn, Robert. "Human aging: usual and successful." *Science*, July, 1987.

Rudman, D., Feller, A.G., Nagraj, H.S., et al. "Effects of human growth hormone in men over 60 years old." *New England Journal of Medicine*, 323(1): 1–6, 1990.

Shetty, K.R. and Duthie, E.H. "Anterior pituitary function and growth hormone use in the elderly." *Endocrinology and Metabolism Clin. North America*. 24: 213–230, 1995.

Yarasheski, K.E. and Zachwieja, J.J. "Growth hormone therapy for the elderly: the fountain of youth proves toxic." *Journal of the American Medical Association*, 270(14): 1694, 1993.

Chapter Four

Crist, D.M., Peake, G.T., Egan, P.A., and Waters, D.L. "Body composition response to exogenous

GH during training in highly conditioned adults." *Journal of Applied Physiology*, 65: 579–584, 1988.

Deacon, James. "Biceps in a Bottle." *MacLean's*, September 7, 1992.

Hakkinen, K. and Pakerinen, A. "Serum hormones and strength development during strength training in middle-aged and elderly males and females." *Acta Physiology Scandinavia*, 150:211, 1994.

Holmes, S.J. and Whitehouse, R.W. "Effect of growth hormone replacement on bone mass in adults with adult onset growth hormone deficiency." *Clinical Endocrinology*, 42: 627–633, 1995.

Kraemer, W.J., Marchitelli, L., Gordon, S.E., Harman, E., Dziados, J.E., Mellow, R., Frykman, P. et al. "Hormonal and growth factor responses to heavy resistance exercise protocols. *Journal of Applied Physiology*, 69: 1442–1450, 1990.

MacLean, D., Kiel, D.P., and Rosen, C.J. " 'Low dose' growth hormone for frail elders stimulates bone turnover in a dose-dependent manner." *Journal of Bone Mineral Research*, 10(Supplement 1): S458, 1995.

Marcus, R., Butterfield, G., Holloway, L., et al. "Effects of short-term administration of recombinant human growth hormone to elderly people." *Journal of Clinical Endocrinology and Metabolism*, 70(2): 519–527, 1990.

Nicklas, B.J. and Ryan, J. "Testosterone, growth hormone, and IGF-1 responses to acute and chronic resistive exercise in men aged 55–70." *International Journal of Sports Medicine*, 16:445–450, 1995.

Rosen, H.N., Chen. V., Cittadini, A., et al. "Treatment with growth hormone and IGF-1 in growing rats increases bone mineral content but not bone mineral density." *Journal of Bone Mineral Research*, 10: 1352–1358, 1995.

Rosen, C.J., and Donahue, L.R. "Insulin-like growth factors: potential therapeutic options for osteoporosis." *Trends in Endocrinological Metabolism*, 6 (7): 235–240, 1995.

Rosen, H.N., Chen, V., Cittadini, A. et al. "Treatment with growth hormone and IGF-1 in growing rats increases bone mineral content but not bone mineral density." *Journal of Bone Mineral Research*, 10: 1352–1358, 1995.

Schwarzenegger, Arnold. "Anabolic steroids and Ergogenic Aids." *The Encyclopedia of Bodybuilding*. New York: Simon & Schuster, 1986.

Smith, R.A., Melmed, S., Sherman, B., Frane, J., et al. "Recombinant growth hormone treatment of amyotrophic lateral sclerosis." *Muscle and Nerve*, 6: 624–633, 1993.

Chapter Five

Barrett-Connor, E. L. "Testosterone and risk factors for cardiovascular disease in men." *Diabetes and Metabolism*, 21: 156–161, 1995.

Cittadini, A., Stromer, H., Katz, S., et al. "Differential cardiac effects of growth hormone and insulin-like growth factor I in the rat." *Circulation*, 93(4): 800–809, 1996.

Fazui, S., Sabatini, D., Capaldo, B., et al. "A preliminary study of growth hormone in the treatment of dilated cardiomyopathy." *The New England Journal of Medicine*, 334 (13): 809–814, 1996.

Jeevanandam, M. and Peterson, S.R. "Altered lipid kinetics in adjuvant recombinant human growth hormone treated trauma patients." *American Journal of Physiology*, 267-E560–565, 1994.

Loh, E. and Swain, J.L. "Growth Hormone for Heart Failure—Cautious Optimism." *New England Journal of Medicine*, 334:E856–857, 1996.

Marin, P. "Testosterone and regional fat distribution." *Obesity Research*, 3 (Supplement 4): 609S–612S, 1995.

Rosen, T., Eden, S., Larson, G., et al. "Cardiovascular risk factors in adult patients with growth hormone deficiency." *Acta Endocrinologica*, 129: 195–200, 1993.

Chapter Six

Baulieu, E.E. "Neurosteroids: A New Function in the Brain." *Biological Cell*, 71: 3–10, 1991.

Gibbs, R.B. "Estrogen and nerve growth factor related systems in brain." *Annals of New York Academy of Science*, 743: 606–612, 1989.

Gouchie, C. and Kimura, D. "The relationship between testosterone levels and cognitive ability patterns." *Psychoneuroendocrinology*, 16:323–334, 1991.

McGauley, G.A., Cuneo, R.C., Saloman, F., et al. "Psychological well-being before and after growth hormone treatment in adults with growth hormone deficiency." *Hormone Research*, 33(Supplement 4): 52–54, 1990.

Rosen, T. "Decreased psychological well-being in adult patients with GH deficiency." *Clinical Endocrinology*, 40: 111–116, 1994.

Sherwin, B.B. "Sex hormones and psychological functioning in postmenopausal women." *Experimental Gerontology*, 29:423–430, 1994.

Chapter Seven

Auernhammer, C.J. "Effects of growth hormone and insulin-like growth factor 1 on the immune system." *European Journal of Endocrinology*, 133: 635–645, 1995.

Gelato, M.C. "Aging and immune function: A possible role for growth hormone." *Hormone Research*, 45:46–49, 1996.

Goya, R.G., Gagnerault, M.C., De Moraes, M.C., et al. "In vivo effects of growth hormone on thymus function in aging mice." *Brain Behavior and Immunology*, 6:341–354, 1992.

Moldofsky, H. "Central nervous system and peripheral immune functions and the sleep-wake system." *Journal of Psychiatry-Neuroscience*, 19: 368–374, 1994.

Mulligan, K., Grunfeld, C., Hellerstein, M.K., Neese, R.A., and Schambelan, M. "Anabolic effects of recombinant human growth hormone in patients with wasting associated with human immunodeficiency virus infection." *Journal of Clinical Endocrinology and Metabolism*, 77(4): 956–962, 1993.

Schmgelan, M. "Studies suggest growth hormone is an effective treatment of wasting syndrome in HIV." *Endocrine News* 21, #2, April, 1996.

Chapter Eight

Auernhammer, C.J. "Effects of growth hormone and insulin-like growth factor I on the immune system." *European Journal of Endocrinology*, 133: 635–645, 1995.

Baker, V.L. "Alternatives to oral estrogen replacement." *Obstetrics and Gynecology*, 21: 297–299, 1994.

Bucchi, L., Hickson, J.F., Pivarnik, J.M., Wolinsky, I., McMahon, J.C., and Turner, S.D. "Ornithine ingestion and growth hormone release in body-builders." *Nutrition Research*, 10: 239–245, 1990.

Cavagnini, F., Invitti, C., Pinto, M., Maraschini, C., DiLandro, A., Dubini, A., and Marelli, A. "Effect of acute and repeated administration of gamma aminobutryric acid (GABA) on growth hormone and prolactin secretion in man." *Acta Endocrinologica*, 93: 149–154.

Ghigo, E., Ceda, G.P., Valcavi, R., Goffi, S., Zini, M., Mucci, M., Valenti, G., et al. "Low doses of either intravenously or orally administered arginine are able to enhance growth hormone response to growth hormone releasing hormone in elderly patients." *Journal of Endocrinological Investigation*, 17(2): 113–117, 1994.

Merimee, T.J., Rabinowitz, D., and Fineberg, S.E. "Argnine-initiated release of human growth hormone." *New England Journal of Medicine*, 280: 1434–1438, 1969.

Muggeo, M., Tiengo, A., Fedele, D., and Crepaldi, G. "Altered control of growth hormone secretion in patients with cirrhosis of the liver." *Archives of Internal Medicine*, 139(10): 1157–1160, 1979.

Visser, J.J. and Hoekman, K. "Arginine supplementation in the prevention and treatment of osteoporosis." *Medical Hypothesis*, 43(5): 339–342, 1994.

Ward, P.S. and Savage, D.C. "Growth hormone responses to sleep, insulin hypoglycaemia, and arginine infusion." *Hormone Research*, 22 (1-2): 7–11, 1985.

Chapter Nine

"Boosting Your Immune System." *The University of California, Berkeley Wellness Letter*, October, 1993.

Goulart, Frances Sheridan. "What's your youth potential?" *Vibrant Life*, p. 19, Jan.-Feb., 1993.

Murphy, P.J., Badia, P., Myers, B.L., Boecker, M.R., and Wright, K.P. "Nonsteroidal anti-inflammatory drugs affect normal sleep patterns in humans." *Physiology and Behavior*, 55 (6): 1023–1026, June, 1994.

Sandyk, R., Anninos, P.A., and Tsaga, N. "Age-related disruption of circadian rhythms: Possible relationship to memory impairment and implications for therapy with magnetic fields." *International Journal of Neuroscience*, 59(4): 259–262, Aug. 1991.

Whitaker, J. "Act now to protect your health." *Health and Healing*, Sept. 1993.

Glossary

Acromegaly—a condition in which bones of the face, jaw, arms, and legs get larger in middle-aged patients. Caused by oversecretion of growth hormone by the pituitary, usually stimulated by a pituitary tumor but sometimes by an overdose of recombinant human growth hormone (HGH).

Adipose—fatty. Tissue made of fat cells arranged in lobes.

Adrenal gland—the organ that sits on top of each kidney and that makes a variety of hormones including the sex hormones (testosterone, estrogen, and progesterone), stress hormones (epinephrine and norepinephrine), and steroid hormones (DHEA), among others.

Aerobic exercise—physical exercise that relies on the intake of oxygen for energy production.

AIDS (Acquired Immune Deficiency Syndrome)—the constellation of infections that best a person whose immune system has been damaged by the Human Immunodeficiency Virus (HIV).

Alzheimer's disease—a brain disease associated with diffuse degeneration of brain cells, occurring mostly in old age. Its cause is as yet unknown. Hormone replacement of estrogen in postmenopausal women may help prevent or delay the disease; the role of HGH, melatonin, and DHEA is still under investigation.

Amino acids—building blocks of protein molecules necessary for every bodily process. The body does not produce essential amino acids, which are necessary for growth and development, but must obtain them through the diet. Nonessential amino acids are those that the body synthesizes itself. Human growth hormone is made up of 191 amino acids.

Anabolic steroid—a drug compound taken from a male hormone (testosterone) or prepared synthetically. Anabolic steroids aid body growth and promote male characteristics.

Anabolism—the process by which bone, muscle, and other tissue build up. Any process that produces energy in which simple substances are converted into more complex matter.

Antibody—a protein produced in the body in response to contact with an *antigen*. An antibody neutralizes the antigen and creates an immunity to that antigen. An integral part of the immune system.

Antigen—any substance recognized as "foreign" or "nonself" by the immune system. In certain disorders, such as rheumatoid arthritis, the immune system mistakenly targets healthy body cells as nonself.

Antioxidant—a chemical molecule that prevents

oxygen from reacting with other compounds to create free radicals. They protect cells from being damaged.

Arginine—an amino acid made by the digestion of proteins.

Arteries—blood vessels that carry oxygenated blood away from the heart to nourish cells throughout the body.

Aspirin—a drug that reduces inflammation and fever. Also known to affect the platelets in the blood to prevent thickening or clotting. Chemical name: acetylsalicylic acid.

Atherosclerosis—a disease of the arteries in which fatty plaques develop on the inner walls, thus narrowing the passageway.

B cells—cells of the immune system that produce *antibodies* to create *immunity* against certain disease.

Blood glucose—glucose, a simple sugar, that circulates in the bloodstream.

Blood pressure—the force of blood on the walls of the arteries. Two levels of blood pressure are measured: the higher, or systolic, pressure occurs each time the heart pushes blood into the vessel. The lower, or diastolic, pressure occurs when the heart is at rest.

Bone marrow—the soft, fatty tissue that fills the cavities of the long bones of the arms and legs; an important site of immune cell formation.

Calorie—the unit in which energy that comes from food is measured.

Cancer—a group of diseases characterized by the uncontrolled growth of abnormal cells in any organ or tissue of the body. Cancer cells may metastasize, or spread to other parts of the body as well.

Carbohydrates—the sugars and starches in food. Carbohydrates are the main source of energy for all body functions and are needed to process other nutrients. Complex carbohydrates are composed of large numbers of sugar molecules joined together, and are found in grains, legumes, and starchy vegetables.

Cardiovascular system—the heart together with the two networks of vessels: arteries and veins. Transports nutrients and oxygen to the tissues and removes waste products.

Catabolism—a chemical process of the body in which energy is released for use in work, energy storage, or heat production. The body breaks down complex substances into simple compounds. See *metabolism*.

Catecholamines—a group of chemicals that work as *neurotransmitters*. The main catecholamines are dopamine, epinephrine, and norepinephrine. They help regulate heart rate, blood pressure, and other functions.

Cholesterol—a fatlike substance found in the brain, nerves, liver, blood, and bile. Synthesized in the liver, cholesterol is essential in a number of body functions. Excess dietary cholesterol contributes to atherosclerosis and heart disease.

Chromosome—a long-chained molecule containing genes and genetic information. Each chromosome is made up of a double strand of DNA.

Chronobiology—the study of internal body rhythms in order to map hormonal, nerve, and immune system cyclical functions. Chronobiologists hope to design hormone replacement and other strategies based on these cycles that will work more effectively and safely than prescriptions and help extend life.

Collagen—a fibrous protein that forms a connective tissue supporting the skin, bone, tendons, and cartilage.

Corticoids—a hormone created by the adrenal glands. Generally speaking, corticoids have powerful effects as anti-inflammatories and are essential for the breakdown of carbohydrates and fats in the body.

Depression—a medical illness marked by feelings of sadness, despair, lack of worth, and hopelessness. The cause of depression can be hereditary and/or stem from a hormonal imbalance, particularly of the neurotransmitters dopamine, norepinephrine, melatonin, and serotonin.

DHEA—a natural steroid hormone secreted by the adrenal glands in youth and early adulthood, and the most common steroid in the blood. Currently investigated for its role in aging and health.

Diabetes mellitus—failure of body cells to use glucose due to a lack of insulin or a failure of the body to use insulin properly.

DNA (deoxyribonucleic acid)—the genetic material of all living things found mainly in the chromosomes in the nucleus of a cell. Along the length of each strand of DNA lie the genes, which contain the genetic material that controls the inheritance of traits.

Dwarfism—the abnormal underdevelopment of the

body with a number of causes, including genetic defects, pituitary or thyroid dysfunction leading to hormone deficiency, chronic diseases, and nutritional problems such as intestinal malabsorption.

Edema—abnormal collection of fluid in tissues, leading to swelling.

Endocrine system—the system of glands and other structures that secrete hormones into the bloodstream, including the adrenals, ovaries, pancreas, pineal, pituitary, testicles, and thyroid.

Estrogen—a group of female hormones responsible for the development of secondary sex traits and aspects of reproduction. Produced in the ovaries, adrenal glands, testicles, and fatty tissue.

Fatty acids—acids found in fats. Essential fatty acids form prostaglandins, which help organ muscles contract, regulate stomach acid, lower blood pressure, and help regulate hormones, among other actions. Saturated fatty acids are found in meat and may lead to high levels of cholesterol if consumed in too high quantities.

Fight-or-flight response—the body's response to perceived danger or stress, involving the release of hormones and subsequent rise in heart rate, blood pressure, and muscle tension.

Free radical—a molecule containing an odd number of electrons, making it highly reactive and, as a result, potentially dangerous to healthy cells.

Gene—the basic unit of heredity. Genes are made of DNA.

Glucagon—a hormone produced in the pancreas that raises the level of blood glucose.

Glucose—the most common simple sugar, also known as dextrose; the chief source of energy in humans.

Glycogen—stored form of sugar in the liver and muscles that is released as glucose when needed by cells for energy.

Hayflick limit—the limit on the number of generations over which a cell may divide. There is a different Hayflick limit for each different kind of cell.

High-density lipoprotein (HDL)—a lipid-carrying protein that transports the so-called "good" cholesterol away from the artery walls to the liver.

Hormone—a chemical produced by the endocrine glands or tissue that, when secreted into body fluids, has a specific effect on other organs and processes. Hormones are often referred to as "chemical messengers," and they influence such diverse activities as growth, sexual development, metabolism, and sleep cycles. Hormones also are instrumental in maintaining the proper internal chemical and fluid balance.

Humanotrope—a brand of human growth hormone.

Human growth hormone—a hormone secreted by the pituitary gland that is instrumental in regulating growth. It is released in periodic bursts during the day and night, especially during sleep, and is controlled by the hypothalamus. HGH promotes protein building in all cells, increases use of fatty acids for energy, and reduces use of carbohydrates. Growth effects depend on the presence of thyroid hormone, insulin, and carbohydrate.

Hyperglycemia—a condition in which the blood glucose is higher than normal.

Hyperinsulinism—too high a level of insulin in the blood. It can occur when the body makes too much insulin, when too much replacement insulin is injected, or when too much human growth hormone flows in the bloodstream.

Hypoglycemia—a condition in which the blood sugar is lower than normal.

Hypopituitarism—a medical condition in which the pituitary fails to release its hormones. This is usually related to a tumor, infection, or surgery.

Hypothalamus—a portion of the brain that activates, controls, and integrates part of the nervous system, the endocrine processes, and many bodily functions, such as temperature, sleep, and appetite.

Immunity—the quality of being highly resistant to a disease or *antigen* after initial exposure and response by the immune system.

Impaired glucose tolerance—blood glucose levels higher than normal, but below the level of diabetes.

Insomnia—a chronic inability to sleep, or to remain asleep, at night. Caused by a variety of factors, including diet and exercise patterns, emotional stress, and hormonal imbalances.

Insulin—a hormone released by the pancreas in response to increased levels of sugar in the blood. Acts to regulate the body's use of sugar and some of the processes involved with fats, carbohydrates, and proteins.

Insulin resistance—a condition in which the body does not use insulin properly; insulin produced by the pancreas is much less effective at moving glucose from the blood into the cells. Some scientists think that rHGH therapy may cause insulin resistance in some patients.

Insulin growth factors 1 (IGF-1) and 2 (IGF-2)—two substances produced in the liver and other tissues in response to hormonal stimulation. IGF-1 works directly with HGH to promote growth and regulate metabolism, while IGF-2 works primarily to maintain the health and function of nerves. The exact role of these growth factors remains under investigation.

Interferon—a group of proteins released by cells that have been infected with a virus. Interferon appears to inhibit viral growth.

Interleukin—any of eight proteins that control aspects of blood cell production and the immune response.

Islets of Langerhans—special groups of cells in the pancreas that make and secrete hormones that help the body break down and use food.

LDL (low-density lipoprotein)—a protein comprised of fats, large amounts of cholesterol, and triglycerides. LDL is the "bad cholesterol" considered a risk factor for the development of heart disease.

Lysine—an essential amino acid needed for proper growth in infants and for maintenance of nitrogen balance in adults. L-lysine supplements may help trigger the release of HGH.

Melatonin—a hormone released into the bloodstream by the pineal gland. Melatonin production is stimulated by darkness and inhibited by light. It is known to act as a sleep promoter, antioxidant, and immune system booster.

Menopause—the end of a woman's fertile life; the loss of estrogen due to ovarian failure.

Metabolism—the sum of all chemical processes that take place in the body to convert food to energy and other substances needed to sustain life. The first step is the constructive phase (anabolism) in which smaller molecules (amino acids) are converted to larger molecules (proteins). The second phase is the destructive phase (catabolism) in which larger molecules (like glycogen, sugar stored in the liver) are converted to smaller molecules (like glucose, blood sugar). Exercise, body temperature, hormone activity, and digestion all affect metabolism.

Muscle—a tissue made up of fibers that are able to contract, allowing movement of the parts and organs of the body.

Neurotransmitter—a chemical that changes or results in the sending of nerve signals. Serotonin, norepinephrine, acetycholine, and dopamine are among the many neurotransmitters that send and receive messages in the brain and body.

Norepinephrine—a stress hormone secreted by the adrenal glands that raises blood pressure and heart rate.

Osteoblast—a cell that works in forming bone tissue during remodeling. Osteoblasts bring together the substances that form the bone.

Osteoclast—a cell that works to break down bone tissue in the process of bone formation called remodeling.

Osteoporosis—the condition in which bones become thin and porous as a result of calcium loss and poor bone metabolism.

Pancreas—the gland that lies behind the stomach and produces the hormone insulin as well as secretes a digestive fluid.

Pineal gland—the hormone gland located in the brain that secretes melatonin. The pineal gland eventually begins to shrink and calcify during the aging process, thereby significantly reducing the amount of circulating melatonin.

Pituitary gland—the small gland joined to the hypothalamus at the base of the brain. It supplies many hormones that control growth, sexual development, and a host of other essential body functions.

Progesterone—a female sex hormone secreted by the adrenal glands and the ovaries. Levels rise during the second phase of the menstrual cycle.

Protein—any of a large group of complex, organic nitrogen compounds. Each is made up of linked amino acids that have the elements carbon, hydrogen, nitrogen, and oxygen. Protein is the main building material for muscles, blood, skin, hair, nails, and organs. It is also needed to form hormones, enzymes, and antibodies.

Protein metabolism—the ways in which protein in foods is used by the body for energy and to make other proteins. Food proteins are first broken down into amino acids, then absorbed into the bloodstream and used in body cells to form new proteins.

Prozac—an antidepressant compound that works to block the reuptake of the neurotransmitter serotonin. Its generic name is fluoxetine hydrochloride.

Recombinant—referring to a cell or organism that results from the rejoining of genes in the DNA molecule. The change can occur naturally or synthetically.

Recombinant DNA—a DNA molecule that has been broken into pieces that are then put back together in a new form. Parts of DNA material from another organism may also be placed into the molecule.

Remodeling—a term used to describe the process by which bone cells die and then are replaced by new cells. It consists of two stages: resorption, in which cells called osteoclasts dissolve some bone tissue to create a small cavity on the bone surface; and formation, in which osteoblasts take nutrients from the bloodstream to fill the cavities.

REM sleep—rapid-eye movement sleep, the stage of sleep during which dreaming takes place. There are generally four to six REM sleep stages per night, each lasting from a few minutes to half an hour.

Serotonin—a naturally occurring neurotransmitter derived from the amino acid tryptophan.

Somatotropin—another name for human growth hormone.

Somatotropin Deficiency Syndrome—a condition in which adults have low levels of human growth hormone.

Testosterone—the primary male sex hormone responsible for male sex characteristics.

Thymus gland—a small endocrine gland located in the upper chest that regulates the development of certain immune system cells and makes hormones important in maintaining a strong, healthy immune system.

Thyroid gland—an organ at the front of the neck responsible for producing thyroxine. The thyroid begins to shrink as we age, slowly decreasing the amount of thyroxine available to the body. Replacing growth hormone may help maintain the integrity of the thyroid.

Thyroxine (thyroid hormone)—a hormone affecting body temperature and the metabolism of protein, fat, and carbohydrates. Thyroxine also keeps up human growth hormone release, skeletal maturation, and the heart function.

Toxin—any substance with the potential to cause disease or damage to body tissues.

Tryptophan—an essential amino acid used to produced both serotonin and melatonin. Found in foods such as legumes, grains, and other sources of protein.

Vitamin—any of a group of substances required by the body for healthy growth, development, and cell repair.

Index

Expertly detailed, pharmaceutical guides
can now be at your fingertips
from U.S. Pharmacopeia

THE USP GUIDE TO MEDICINES
78092-5/$6.99 US/$8.99 Can

- More than 2,000 entries for both prescription
  and non-prescription drugs
- Handsomely detailed color insert

THE USP GUIDE
TO HEART MEDICINES
78094-1/$6.99 US/$8.99 Can

- Side effects and proper dosages for over 400
  brand-name and generic drugs
- Breakdown of heart ailments such as angina,
  high cholesterol and high blood pressure

THE USP GUIDE TO
VITAMINS AND MINERALS
78093-3/$6.99 US/$8.99 Can

- Precautions for children, senior citizens and
  pregnant women
- Latest findings and benefits of dietary supplements

THE NATIONWIDE #1 BESTSELLER

the Relaxation Response

by Herbert Benson, M.D.
with Miriam Z. Klipper

A SIMPLE MEDITATIVE TECHNIQUE
THAT HAS HELPED MILLIONS
TO COPE WITH
FATIGUE, ANXIETY AND STRESS

Available Now—
00676-6/ $6.99 US/ $8.99 Can